The World University Library is an international series of books, each of which has been specially commissioned. The authors are leading scientists and scholars from all over the world who, in an age of increasing specialization, see the need for a broad, up-to-date presentation of their subject. The aim is to provide authoritative introductory books for university students which will be of interest also to the general reader. Publication of the series takes place in Britain, France, Germany, Holland, Italy, Spain, Sweden and the United States.

J. G. Baer

Animal Parasites

Translated from the French by Kathleen Lyons

World University Library

McGraw-Hill Book Company
New York Toronto

Library of Congress Catalog Card Number: 70-118404

Photoset by BAS Printers Limited, Wallop, Hampshire, England
Manufactured by LIBREX, Italy

Contents

multiply ye there,
and be not diminished.
JEREMIAH 29:6

1 The parasitic way of life

Since the earliest times parasites have intrigued physicians and naturalists, who were astonished to find organisms living in the intestine or on the surface of the body of human and domestic animals. Numerous theories were elaborated to account for their origin. Some preferred to think that parasites arose by spontaneous generation in the intestine or in the detritus and dirt on the outside of the body; others, seeking to reconcile their ideas with the biblical account in Genesis, believed them to have been created in Adam and their 'germ' transferred to Eve with the rib. Parasites are in fact more widespread than one might suppose, occurring in nearly all vertebrates as well as in numerous invertebrates. Indeed it is rare to find organisms that are not parasitised or to find classes of invertebrates which have not adopted this way of life. Parasitism is one of those aspects of biology and, in particular, ecology, which deals with the relationships of organisms to one another and to their habitat. The study of parasites is complicated by the ways in which they are adapted to their habitat – that is, to their hosts – for they may often show considerable morphological and physiological specialisation.

Many kinds of association, which may be more or less tenuous, are known to exist between individuals of different species, but the true nature of these relationships cannot be understood properly without more experimental investigation. Despite this lack of information there is a tendency to include all associations of this kind under the term *symbiosis* (Henry, 1966–7); while perhaps this is correct from the etymological point of view (symbiosis was first used as a general term meaning merely the living together of two organisms of different species) it nevertheless masks the real biological nature of these relationships.

Many of these associations have developed from chance meetings in the same biotope and they may offer certain advantages to one of the partners. For example, the sedentary vorticellids which attach themselves to the surface of fresh water insects and crustaceans benefit from the movements of their hosts, which help to renew the supply of oxygen and to provide food. This kind of association, termed *phoresy*, is not usually obligatory for these protozoans because they also attach to inert objects.

Wood-eating insects and herbivorous mammals that are essentially cellulose feeders harbour in their guts an abundance of protozoans, bacteria and yeasts and these organisms are able to digest cellulose or to manufacture amino acids. When experimentally deprived of these organisms they waste away, despite the fact that they continue to feed, because they lack the necessary enzymes to degrade cellulose. On the other hand, the cellulose-degrading organisms seem to have specific requirements similar to those found in the gut of the normal host, for they are difficult to culture on artificial media. It seems that in this kind of association, called symbiosis, the two partners are mutually dependent. A size difference between the two partners is usual in animal associations of this kind. In plants the participants are often of about the same size and the association may result in a new organism, as is true of the lichens, which are constituted by an association between an alga and a fungus. The term symbiosis defined in this restricted sense can, therefore, be considered as a kind of reciprocal parasitism where each partner obtains from the other essential food materials which it lacks.

Parasitism, in contrast, involves an association between animals of different species where one, the host, is indispensable to the other, the parasite, while the host can quite well do without the parasite. From the stage when the parasite becomes completely dependent on its host, the host-parasite association becomes a biological unit which may be permanent as far as ecological conditions in the host environment allow.

Some authors employ a more restricted view and consider as

1·1 Above A piece of skin from the grey whale *Eschrichtius gibbosus* on which are specimens of the phoretic barnacle *Coronula diadema.* The dark ring is formed by the host epidermis, which pushes between the plates of the barnacle and anchors it in position. In addition, all stages in the life history of the amphipod *Cyamus scammoni,* a micropredator of the whale, are shown.

1·2 Right The phoretic protozoan *Ellobiophrya donacis* on the gill of *Donax,* the banded wedge shell.

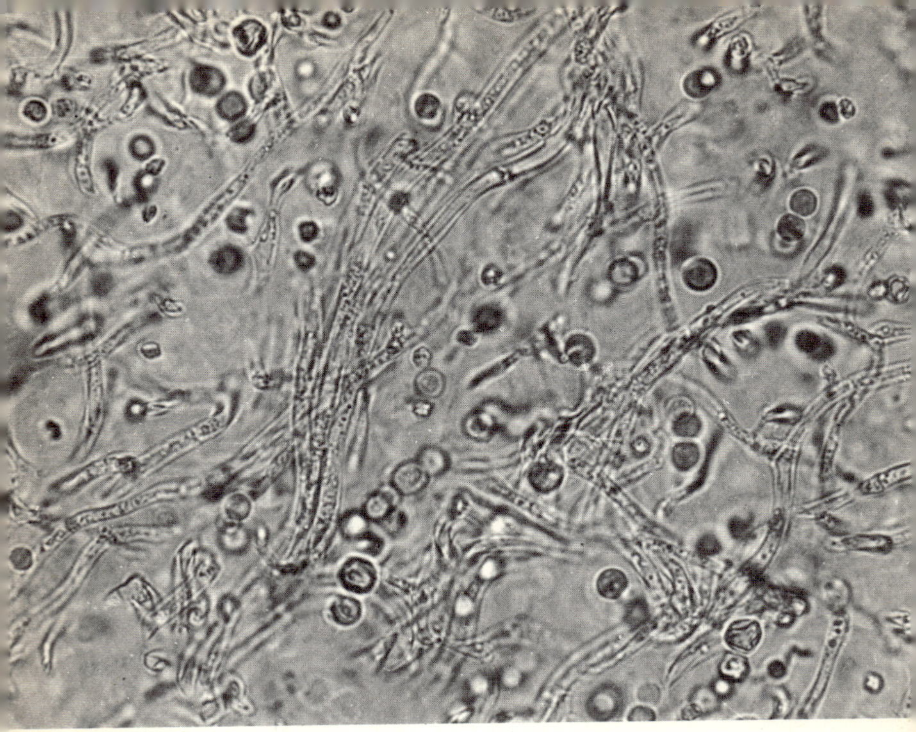

parasites only those organisms living deep within the tissues of the host and in consequence evoking immunological reactions (Sprent, 1963) but this restriction would lead to the paradox that only larval *taeniids* were truly parasites while the adults inhabiting the intestine were not.

Normally, parasites never kill their hosts, for this would be equivalent to depriving themselves of food and committing suicide. However when the host is subjected to unfavourable conditions, for instance during illness or while in captivity, its parasites may prove sufficiently injurious to kill the host. The Okapi, which lives in the tropical forests of central Africa, harbours at least five kinds of worms simultaneously and some of these may be present in numbers of several hundreds; the host does not seem any the worse for this and can feed itself as well as cater for the fauna it contains, but when taken into captivity and fed in the same

1·3 Opposite A transverse section through a lichen. The filaments are fungi and the round, separate cells are algae.
1·4 Examples of high levels of infestation in healthy hosts. **Below** *Anisakis typica* in the stomach of a seal. **Right** *Taenia crassiceps* (proliferative larva) in a fox.

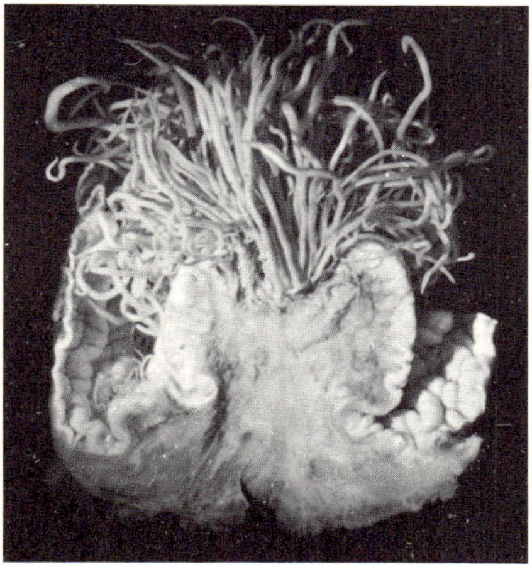

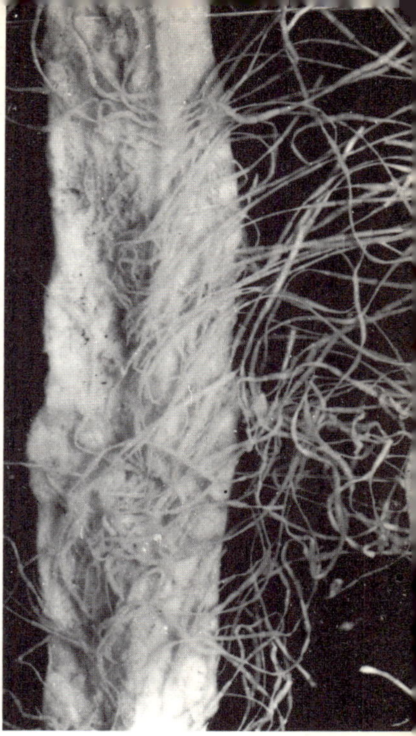

way as ordinary ruminants it often cannot compensate for the drain imposed by the parasites. It weakens and finally dies.

Parasites must also be distinguished from predators which, in general, kill their prey. Predators are normally thought to be larger than their prey, for example the lion and its prey the wart hog, fox and rodents, barn-owl and mice. In fact there are many micropredators, sometimes called *parasitoids*, which are smaller than their victims which they usually enter at the egg stage. The larvae of many hymenopteran and dipteran insects thus eat other developing insects from inside and invariably kill them. They are micropredators although the adult insects are free-living. This peculiarity, common to a number of insects, of having predatory larvae, has been used to control economically undesirable insects. This method is termed biological control and has the advantage that it does not injure other organisms, as do chemical pesticides.

Parasitism, as understood here, is more widespread among plants and animals than one might suppose. Examples used here to illustrate the evolution of parasitism and the nature of the host-parasite relationship are taken from the animal kingdom as the structure of plants makes their approach to parasitism very different, making a comparative account hardly possible.

Parasites do not form a morphologically homogenous group because this way of life has appeared several times in the course of invertebrate evolution. In vertebrates, however, the phenomenon of parasitism is rather like a kind of interspecific graft and not comparable to an association between individuals of different species. For example, there are certain parallels with the development of the mammalian foetus *in utero* and the case of some male fish which 'graft' themselves on to the female.

From the biological point of view, the nature of the parasite biotope must be understood before the adaptive morphological changes involved in parasitism become meaningful.

While many forms are parasitic throughout all stages of their existence, others have adopted this way of life only in the larval stage, the adults being free-living; conversely, they may be parasites as adults and free-living as larvae. Several examples of life cycles are given in table 1 (p. 13) and it is evident that what is most important physiologically is the transition between a free-living and parasitic way of life and *vice versa*. Fleas, for example, have larvae looking like small caterpillars which have mouthparts that allow them to chew food composed of detritus and fungus growing in the nest or den of the host. After metamorphosis, however, the adult flea has piercing and sucking mouthparts which force it to seek its specialised diet of host blood. Several fly larvae inhabit the tissues of their hosts and feed within natural cavities, the nose, stomach and intestine. When metamorphosis is about to occur they leave the host, become soil-dwelling and pupate. The adult fly,

Table 1

Group of parasites	Larvae	adults
anodontids	parasitic	free-living
gastropods	free-living	parasitic
cirripedes	free-living	parasitic
copepods (monstrillids)	parasitic	free-living
nematodes	free-living	parasitic
	parasitic	parasitic
cestodes	parasitic	parasitic

which emerges in spring, has rudimentary mouthparts and does not feed, reserves accumulated during larval life sufficing to allow fertilisation of the female and successful egg production. In the strepsipterans, which parasitise other insects, especially solitary bees, the winged adult males are free-living while the wingless females and the larvae of both sexes are parasites.

These examples illustrate the way in which the classification of these many kinds of relationship can easily become artificial. Despite this they are still of some value in that they emphasise the difficulties encountered in trying to group into distinct categories biological phenomena which have no defined limits.

Although parasitism appears under many different guises, one can nevertheless find a number of characters common to this way of life. It would be a mistake to adopt the commonly held view that parasites are degenerate instead of seeing them as specialised organisms, closely adapted to a particular biotope and with an enhanced reproductive potential which ensures their survival. Many parasites are indeed greatly modified, and when on occasions it is possible to compare them with their nearest free-living

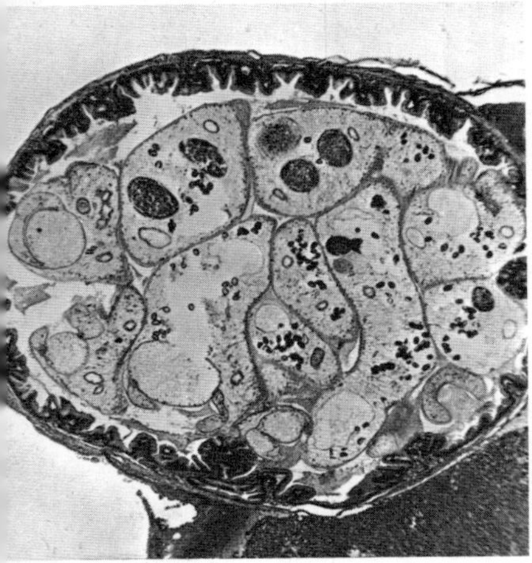
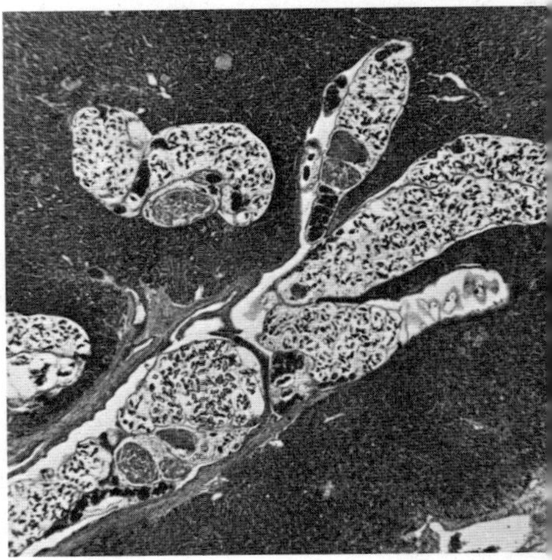

neighbours it is found that modifications of now redundant locomotory organs or hypertrophy of particular structures may be associated with special functions such as nutrition of the parasite or incubation of its eggs. The risk that a parasite may not find a suitable host is compensated for by an often spectacular increase in egg output or by the appearance of a process of larval multiplication during development of the parasite. Parasitism has not resulted in the production of new organs but has led to the adaptation of previously existing organs to suit new ends. During the development of certain ectoparasitic crustaceans, for example, the second maxilla is transformed into a sucker at the end of a moult (figure 1·8). Organs such as suckers or hooks, by means of which the parasite maintains itself on the interior or exterior of its host, are particularly well developed, and owing to the selective advantage they confer have become greatly diversified during the course of evolution. It must be remembered, however, that such adhesive organs are not exclusively a feature of parasites; suckers occur, for instance, in free-living animals such as the cephalopods, which use them to capture prey, and even in some fish which attach themselves to rocks or to other fish. The anatomical principle of the sucker as an attachment organ has always carried a selective advantage and it is not surprising that it also favoured the evolution of parasitism.

1·5 Opposite left A section of the gall bladder of a shrew containing many specimens of the trematode *Dicrocoelium soricis.* **Opposite** A section through the pancreas of a blackbird showing the ducts full of the trematode *Euamphimerus pancreaticus.*

1·6 Right A section through the intestine of a guillemot showing two scolices of *Anomotaenia meinertzhageni* deeply buried in the wall.

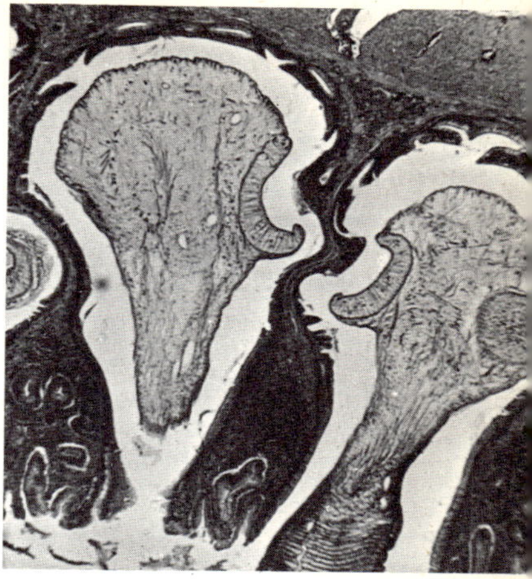

Wherever free-living and parasitic forms occur within the same group of animals it is usually possible to see what kind of morphological adaptations the latter have undergone relative to the free-living forms. In the case of the nematodes or round worms, however, which is a large group containing more than a thousand species, it is very difficult to distinguish either fundamental morphological or physiological differences between the parasitic and the free-living forms. It is therefore considered that the nematodes are preadapted to parasitism and that the transition from an independent to a parasitic existence has occurred on several occasions from different, free-living, groups (Chabaud, 1965).

The life cycles of parasites are often complicated by the existence of successive larval forms which are themselves parasites of different hosts. The host harbouring the adult parasite is termed the *definitive* host and that necessary for the development of a particular larval stage is termed the *intermediate* host. It follows that, where there are several successive larval stages in a life cycle, there are usually as many intermediate hosts as there are larval stages. It is possible that the last intermediate host harbouring the final larval stage may be ingested by an unfavourable host instead of being eaten by the normal definitive host. This gives rise to two possibilities: either the larva is digested and the cycle broken or the larva

1·7 The larva of the fly *Pollenia rudis* is a micropredator; it enters earthworms by the genital pore and eventually destroys the worm completely.

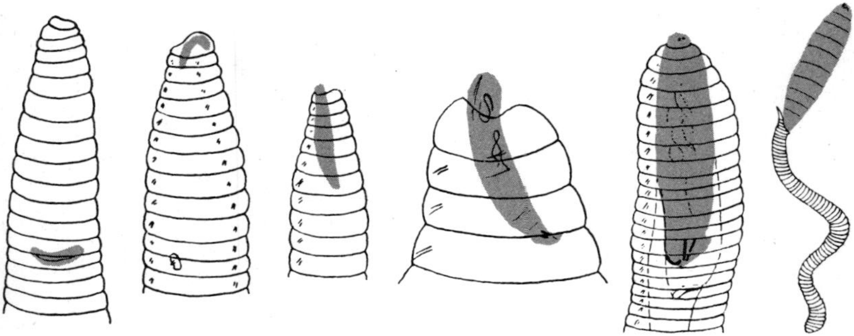

manages to penetrate the intestine and enter peripheral tissues, where it lives without undergoing further development. This host, which is not usually necessary to the life cycle, can nevertheless aid its completion and even lead to the acquisition of new definitive hosts; this host is termed the *paratenic* host.

The successful completion of a life cycle therefore depends on the fact that the intermediate and definitive hosts occur in the same ecosystem or, often, in a particular niche in that ecosystem. This situation favours biological isolation, which leads to specialisation of the life cycle and of the morphology and physiology of the parasites. Isolation of adult and larval parasites on their hosts has given rise to the phenomenon called *host specificity*, which is dealt with in chapter 8.

In parasitic Protozoa it is often not possible to recognise intermediate or definitive hosts and therefore the term *vector* is used to qualify the host responsible for transmitting the parasites to new hosts.

The parasitic way of life demands physiological specialisation to the particular milieu inside or on the exterior of the host. Despite recent studies on the biochemistry and ultrastructure of parasites, their precise methods of feeding and metabolic requirements are not yet fully understood. It has been clearly demonstrated, however, that results obtained with one parasite cannot be

1·8 The replacement of a claw by a sucker at the end of a moult in a crustacean parasitic on fish.

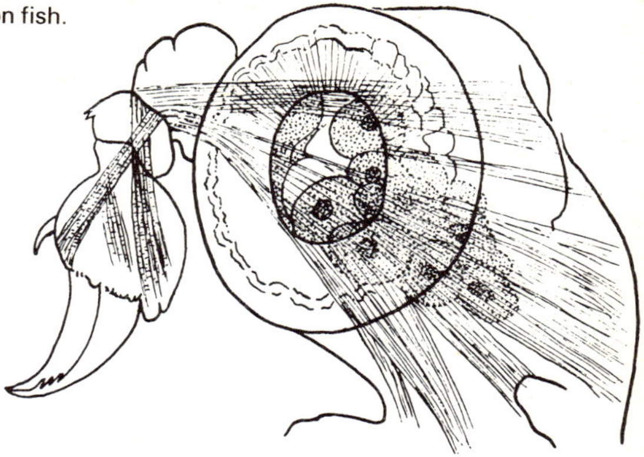

extrapolated to others, so generalisations are premature at the moment. The problem is also complicated by the difficulty of obtaining sufficient material from all stages in the life cycle without having to turn to the ordinary parasites of domestic animals, which are often the least representative of the group. It is to be hoped that the development of micromethods will give parasitology new analytical techniques which will further our understanding of the biochemistry of host-parasite relationships and will assist in uncovering the causes and origins of parasitism which at present remain highly speculative.

2 Adaptations to parasitism

In this chapter it will be shown how free-living organisms have been able to adapt themselves to parasitism and how they have undergone transformations which, although often changing them unrecognisably, have nevertheless assured the success of this new enterprise.

The Protozoa will be treated separately since they cannot really be compared with an isolated metazoan cell. Parasitism has appeared in the Protozoa on several occasions and some groups like the Sporozoa are exclusively parasites whilst other groups have only a few parasitic members. In some instances it is difficult to gauge whether the relationship is truly parasitic or whether it is saprophagy or symbiosis. Amoebae found in the human intestine are classed as parasites but are really ordinary saprophagous species feeding on bacteria and yeasts which occur in the intestine. On the other hand, *Entamoeba histolytica*, the cause of amoebiasis and amoebic dysentery, is a pathogen which attacks and destroys the intestinal lining and feeds on red blood cells, though it can also exist as a harmless form feeding on bacteria. So far there has been no satisfactory explanation for this unusual situation. Experimental work suggests that the intestinal contents affect the amoeba in some way rather than that there are distinct races of amoeba, one being a pathogen and the other harmless (Elsdon Dew, 1968).

Flagellates are considered to be the oldest group of the Protozoa, and that which lies at the base of the animal and vegetable kingdoms. Many flagellates are commensal or symbiotic in the gut of wood-eating insects, but here we will concentrate on the principal representatives of the family Trypanosomatidae. All the species are parasitic and their relationships with their hosts may provide clues as to how parasitism could have arisen.

The trypanosomes must have arisen from forms living in the gut of insects and perhaps even in those aquatic invertebrates where the entire life cycle takes place within the gut. They are designated as monogenetic, and are now grouped into four genera according to ultrastructural characters. They show, during the course of their development in the insect, successive stages in which the flagellum and the kinetoplast are either anterior to the nucleus (promastigote) or posterior to the nucleus (opisthomastigote). Only in *Blastocrithidia* is there a rudimentary undulating membrane (epimastigote). All these genera have a leishmanial phase, which is a rounded form devoid of a flagellum (amastigote) and evacuated together with the faeces. This leishmanial form corresponds with the infestive stage eaten by other insects. Transmission of the parasites is, then, by direct contamination (Hoare, 1967).

Trypanosomes occurring in the blood of vertebrates have gained access to this medium via blood-feeding insects which harbour the stages mentioned above in their guts. In this case, however, the infestive form is not represented by amastigote organisms but by metatrypanosomes (trypanomastigotes). Two modes of transmission of the metatrypanosomes can occur, according to whether they are formed in the posterior or in the anterior portion of the insect's gut. In the former case, the metatrypanosomes are evacuated with the faeces when the insects feed on a mucous membrane or on a wound and the flagellates penetrate directly into the tissues. This mode of transmission is known as stercoral. In the second case the metatrypanosomes enter the salivary glands and are thus inoculated by the insect. This is known as salivary transmission.

These two modes of transmission correspond to two distinct biological forms of parasitism. The species transmitted by the faeces are not pathogenic and their multiplication in the blood of the vertebrate host is limited or sometimes discontinuous. *Trypanosoma lewisi*, for instance, a parasite of the rat, is transmitted by flea faeces which contain the metatrypanosomes and which the rat ingests when it licks itself. In the host's blood the trypanosomes multiply for only a few days and disappear completely within a few weeks. This also occurs in *Schizotrypanum cruzi*, which is transmitted through the faeces of reduvid bugs to a considerable number of South American mammals and which causes Chagas' disease in man. The cycle differs from the one mentioned above in that the trypanosome stage is preceded by an intracellular amastigote stage.

All the species transmitted via the salivary route are pathogenic and multiply in the host's blood. It is such species which cause nagana, or African cattle trypanosomiasis, and sleeping sickness in man. The geographical distribution of these species coincides with that of the tsetse flies (*Glossina*), which are the only known vectors. It is likely that man became infested from strains originating in cattle and that the latter originally acquired their infection from wild ruminants in which there are no pathological symptoms, this indicating a very ancient association. Repeated passages of *brucei* group polymorphic trypanosomes through laboratory animals bring about changes or select for monomorphic types which are no longer infective to tsetse flies. It is also found that human trypanosomes cultivated in a suitable medium multiply or are inhibited according to the particular human blood added to the culture medium.

From a purely formal point of view, and without taking into consideration medical or veterinary aspects, all pathogenic trypanosomes transmitted by tsetse flies are more or less individualised races of *T. brucei* from cattle. It is likely that the *brucei* strain has given rise to *T. evansi* which causes surra, a disease which occurs in several species of mammals but especially in camels and horses. This form of trypanosome is transmitted directly by horse flies and its geographical distribution extends from north of the tsetse belt in Africa into Asia. According to Hoare (1967), *T. equiperdum*, which causes dourine, a disease of horses transmitted by contact of the male and female mucosae during coition, appears to be derived from *T. evansi*.

It is therefore possible to see how parasitism has evolved among trypanosomes and to what extent ecological and biochemical factors play an important part. It is moreover probable that extended research on the nonpathogenic forms from wild animals that are of no direct medical or veterinary importance would furnish results which could be interpreted more easily. It is interesting to find, for instance, that *T. grayi* from the blood of an African crocodile is also transmitted by tsetse flies but that in this case transmission is of the stercoral type. In South America there occurs in many species of mammals, including man, a nonpathogenic form known as *T. rangeli* whose vector is a reduvid bug. Transmission occurs stercorally but it is also found that a few metatrypanosomes pass through the wall of the gut into the salivary glands. However, this extra-intestinal cycle is far from constant and appears to be dependent on the species or even on the race of the vector.

All sporozoans are parasites either of invertebrates or vertebrates, where they occur in natural cavities such as the intestine or within cells. The intracellular species destroy cells and progressively invade new cells so that only the host's natural or acquired resistance will enable it to survive. This type of parasitism is accompanied by a comparatively high mortality which affects both man and domestic animals. This badly adjusted relationship raises the question of whether the part of the parasite life cycle in the host may not be of secondary importance, since spores can survive outside the host or in the vector until they are eaten by or inoculated into a new host. Sporozoans do show physiological adaptations to their hosts, however, this relationship in some ways resembling true host specificity but without there being any phylogenetic or ecological affinities.

2·1 The sporozoan life cycle.

The adaptation of sporozoans to parasitism is shown to some extent by their life cycles. The latter are always complicated by the presence of a sexual phase (gamogony) followed by an asexual phase (sporogony). In most species, however, an asexual multiplication (schizogony) occurs between these two phases and this plays an important role when invading a new host. The sporozoan life cycle is represented in figure 2·1.

In gregarines, which are essentially parasites of arthropods, the spores are voided into the environment and swallowed by a new host. In the gut of its new host the sporozoite undergoes a period of growth which is followed by gamogony. The entire cycle thus occurs in the gut of a single host but there is no schizogony.

There are two types of life cycle in the Coccidia, which are intracellular parasites of invertebrates and vertebrates. In the first type, as represented by *Eimeria*, all three phases occur in the same host and only the spores are voided. In the second type, as in *Aggregata* for instance, there are two successive hosts; schizogony occurs in a crab whereas gamogony and sporogony are found in cuttlefish which feed on crabs. The spores are eliminated from the cuttlefish's gut and swallowed by crabs.

In Haemosporidia all phases of the life cycle occur within one host. A blood-sucking arthropod, usually a dipteran, inoculates the sporozoites into the vertebrate host. The first stage of schizogony take place in the endothelial cells of the liver and from there the schizozoites invade the red blood cells in which gamogony occurs. Fusion of the gametes followed by sporogony takes place in the gut and the haemocoel of the vector and the sporozoites accumulate in the latter's salivary glands.

Haemosporidia are found in all vertebrates except fishes and most amphibians. Table 2 lists the three principal genera together with their hosts and vectors.

The genus *Plasmodium* includes species causing human malaria.

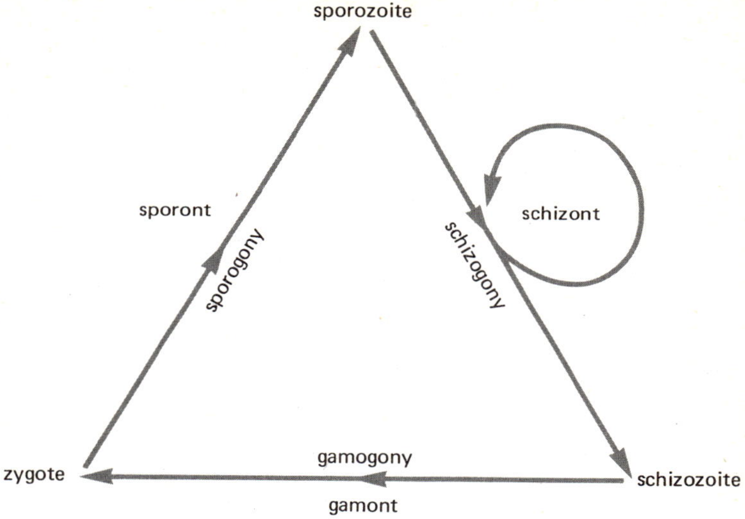

In Brazil and Malaya some species normally living in monkeys also occur in man. Man is also frequently accidentally infested with *P. cynomolgi*, a species occurring in primates which is maintained for research purposes in many laboratories. In fact some six species of plasmodia from monkeys have been transmitted successfully through mosquitoes to human volunteers. Further research may show that several of these species are just races or ecotypes whose study may well help an understanding of the epidemiology of malaria and of the principles of preventive measures.

Babesia are a typical Haemosporidia parasitic, for instance, in the red blood cells of cattle in which schizogony is replaced by binary fission. Sporogony occurs in the ovaries and in the eggs of the tick vector and the sporozoites invade the larvae developing within the eggs. These tick larvae are therefore able to transmit parasites at their first blood meal.

We have seen above that parasitic flagellates were originally intestinal parasites of invertebrates, mainly arthropods. In sporozoans, however, the situation appears to be more complicated, largely because many life cycles remain to be elucidated. For

Table 2 The main genera of Haemosporidia

Genus	Host	Vector
Haemoproteus	Birds	Pupipara, Fleas
	Snakes	Ticks
	Turtles	Leeches
Leucocytozoon	Birds	Blackflies
Plasmodium	Reptiles	Culicid mosquitoes
	Birds	,,
	Mammals	,,

instance, in the genera *Lankesterella* and *Schellackia* the schizozoites penetrate into the red blood cells of frogs and lizards respectively. In *Lankesterella*, schizogony, gamogony and sporogony occur in the endothelial cells of the blood capillaries and the spores burst into these blood vessels, liberating the sporozoites into the blood. The vector is a leech in which the sporozoites are not digested but passed on to another host at the next blood meal. In the genus *Schellackia*, schizogony and gamogony take place within the intestinal epithelum, whereas sporogony occurs in the submucosa where the sporozoites escape to enter the red blood cells. The vector is a mite in which the schizozoites remain unaltered and gain access to another host when the latter eats infected mites.

In both cases the vector plays a purely mechanical role since all the phases of the life cycle take place in the vertebrate host. It is possible that for sporozoan life cycles, particularly the intracellular species, the vector is a secondary acquisition and that vertebrates were the original hosts.

The molluscan and crustacean phyla essentially contain free-living forms but have given rise to parasites on several occasions. As these phyla are represented by complicated and highly differentiated organisms, the comparison with the parasitic species is even more impressive and to some extent allows the adaptation to this new mode of life to be followed.

Parasitic molluscs

Parasitism has appeared in two distinct groups of molluscs and has assumed two different forms because in the unionids and mutelids only the larvae are parasitic, the adults being free-living, whilst in the prosobranch gastropods the larvae are free-living and the adults are parasitic.

Larval parasitism: Larval parasitism occurs in fresh-water mussels of the mutelid and unionid families. Eggs are produced in often enormous numbers: a single *Anodonta*, for example, produces up to two million. Instead of being laid the eggs are retained by the females and are incubated until hatched on the gills, which are modified as a brood chamber. The larvae are ejected in spurts of several dozen at a time by means of the exhalent siphon.

The larvae of *Anodonta*, termed glochidia, have two hardened valves which enclose the embryonic structures. They articulate at the base and can be shut by the contraction of an adductor muscle which joins them. Glands situated in the embryonic mass secrete a substance which runs between the two valves and hardens to form a long byssus thread (figure 2·3). At the apex of each valve is a small articulated spine which folds inside when the two valves are closed. The glochidia lie tangled together by their byssus threads on the mud's surface where the smallest disturbance, caused for instance by the passage of a fish, lifts the bunches of glochidia and brings them into contact with the fish. Stimulated by mucus secreted by the fish's skin the glochidia become very active and the adductor muscle contracts rapidly, enabling some larvae to attach themselves to the fins where they are literally hooked on by the two spines which become buried in the skin. Larvae which lack spines are in general taken in by the respiratory current and attached to the gills of the fish.

2·2 Glochidia larvae of *Anodonta* attached to the fins of a fish.

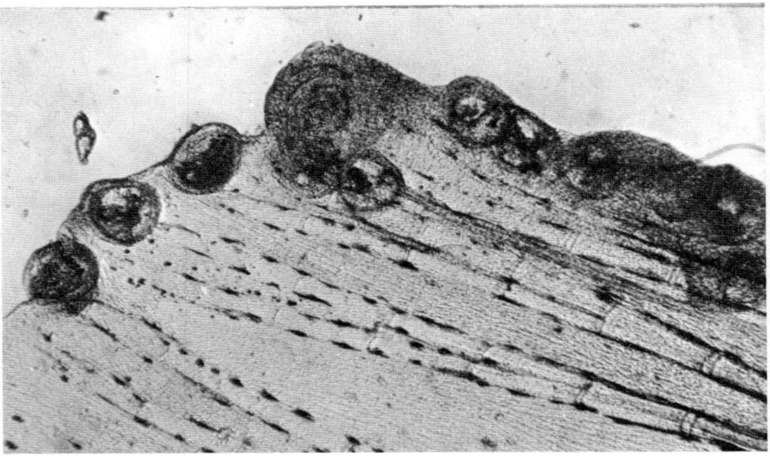

As soon as the glochidia are attached they become immobile, but their presence irritates the skin of the fish which reacts by enveloping each glochidia in a minute cutaneous cyst visible to the naked eye and found particularly on the fins (figure 2·2).

From now on the glochidium will live as a temporary parasite on the fish, undergoing a metamorphosis within the cyst which will transform it into a small mussel. During this period the larva feeds partly on the distintegrating tissues of the host but also on its own tissues which are liquified and serve to nourish the rudimentary embryonic cells of the future mussel. When this has been constituted, it begins to move around within the cyst and eventually breaks the wall, using movements of its foot. The parasitism of larval mussels is obligatory, and out of the enormous number of glochidia produced only a fraction become attached and develop into adult mussels, even though fish carrying large numbers of cysts on their fins are sometimes found. The relationship established between the parasite and its host is examined in chapter 8.

In an African mutelid, *Mutela bourguinati*, there are no glochidia but many thousands of small larvae, each enclosed within a chitinous larval shell which is folded over, and the edges fused

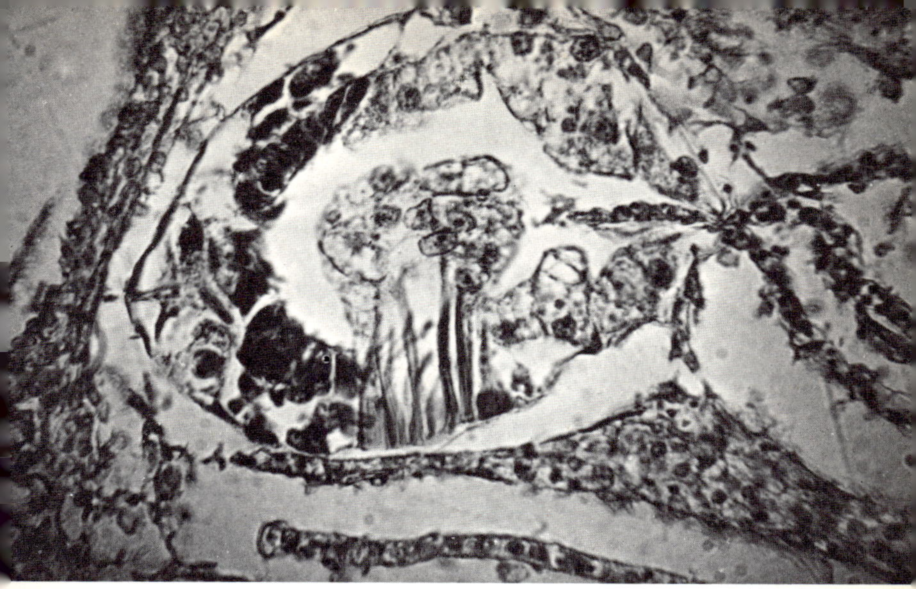

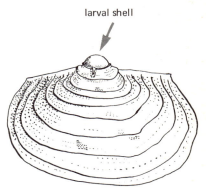

2·3 **Above** A section through an encysted glochidium in the process of metamorphosis.
Bottom left A glochidium.
Bottom right A young mussel with the larval shell in addition to the definitive shell.

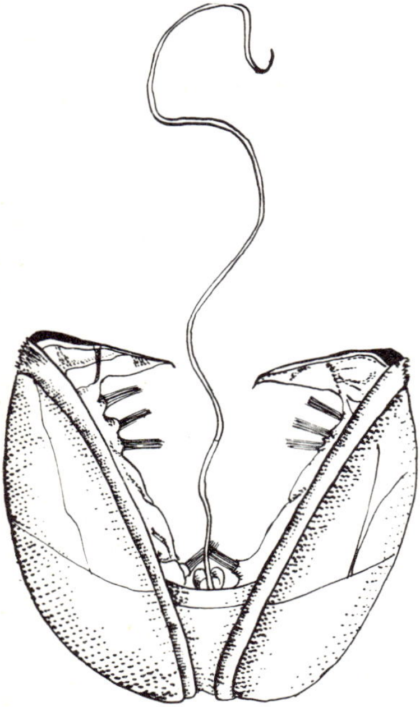

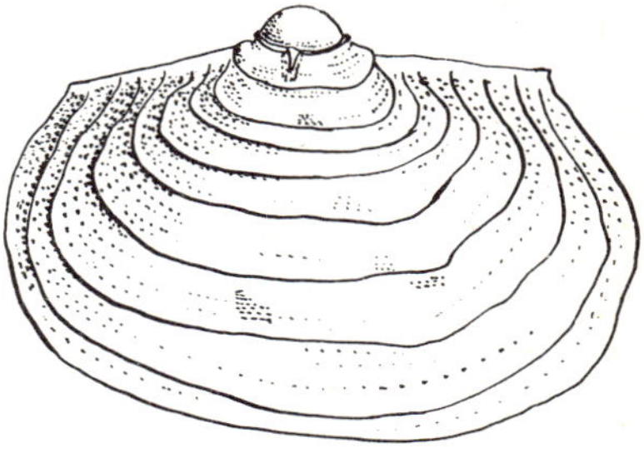

2·4 Lasidium larva of the bivalve *Mutela bourguinati*. (a) Larva hatching from the egg with its filament extended; (b – d) development of the suckers which become buried in the tissues of the fish; and (e) a young mutelid about to become detached at the point indicated by the dotted line.

together, except in the anterior region. Here two small patches of cilia occur, as well as a hollow filament which is at least sixty times as long as the larva. By means of this filament the larva liberated in the river attaches to inert objects and is not swept away by the current. Having made contact with a suitable fish, a barb, the larvae attach themselves to the caudal fin by means of small hooks situated on their ventral surfaces which cause irritation and probably stimulate the fish skin to produce mucus, which coats the larvae. From this point onwards the larvae become parasites and drive two tentacles derived from the larval mantle into the skin of the host; these will grow and serve both to anchor the larva to the host and furnish it with the necessary food for metamorphosis. The larval tissues are resorbed and at the end of a stalk-like extension of the tentacles appears a rudiment enclosing tissue which will form the small mussel. When this is formed the mussel escapes and falls to the bottom while the stalk and tentacles remain in the skin of the host, to be eliminated later (Fryer, 1961).

As the reproduction of fresh-water mussels is seasonal, the inherent risks of this kind of cycle, which can be completed only by the interception of a fish, are compensated for by the enormous mass of larvae produced, as well as by the several years of productive adult life spent by the mussels.

Incubation of eggs in branchial pockets occurs only in fresh-water lamellibranchs, but in one species, not dealt with here, metamorphosis is already completed within the egg when this is laid. In this case the number of eggs is never very large, only a matter of dozens, and the longevity of this species would not be more than a year. Thus the parasitism of incompletely formed larvae compensates for the slow growth of unionids and favours survival of the species.

Adult parasitism: There are about twenty genera of prosobranch

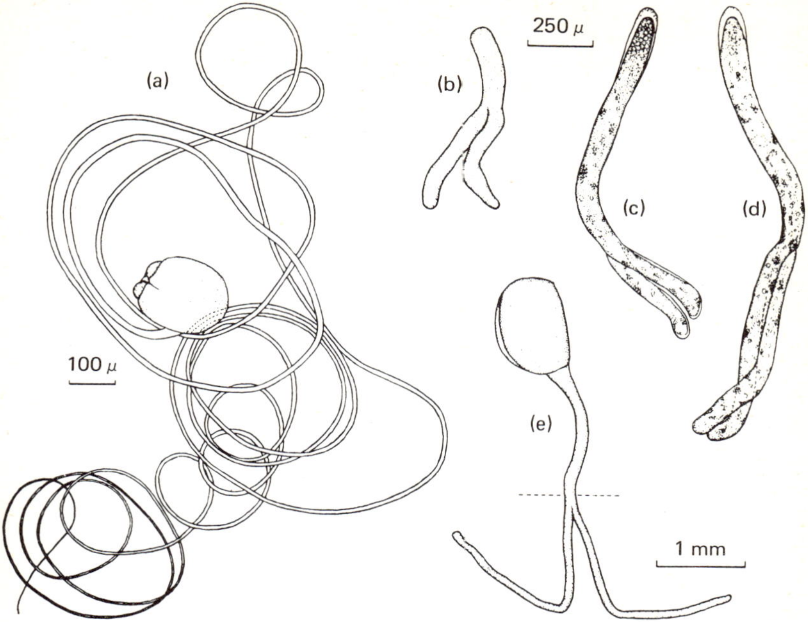

gastropods, all aglossids, which are parasitic as adults. As this group is composed only of marine forms, the hosts are also all marine. Although many of these parasitic forms have been little investigated it is possible to group them as predators, ectoparasites and endoparasites. Within this apparently continuous evolutionary series, however, there occur at least four distinct methods of incubating the eggs, which indicates the presence of different evolutionary lines.

The pyramidellids are predators possessing a long-armed proboscis by means of which they perforate the tissues of their victims, such as sedentary annelids and bivalve molluscs which are always larger than themselves. As the larvae of these predators have only a very short planktonic existence, if indeed this is not completely absent, the young molluscs are hardly separated from their future victims and this favours a strict ecological association. For example, *Odostomia scalaris* is always associated with *Mytilus edulis*, the edible mussel, whilst *O. unidentata* is a predator of the

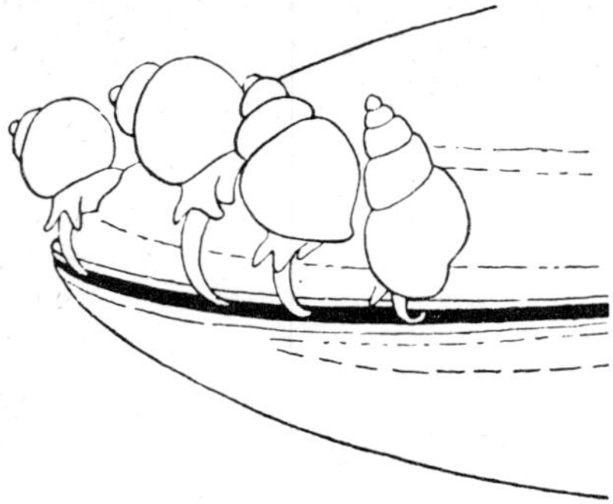

2·5 *Odostomia scalaris*, a bivalve predator.

sedentary polychaete *Pomatoceros triqueter*. *O. eulimoides*, on the other hand, is associated with many kinds of bivalve molluscs, such as scallops and oysters.

All the truly parasitic gastropods are associated with echinoderms, mainly with holothurians but more rarely with sea urchins or star fish, which, living partially buried in the sand, form particularly accessible hosts for a mollusc. Some are ectoparasites which attach themselves either to the surface of their host or bury into it to varying extents; others are endoparasites which live inside the body of the host. Certain kinds of ectoparasites which although deeply buried in host tissue remain in communication with ambient sea water, constitute a special case. Here use of the term ectoparasite implies that in general the parasite has penetrated via the surface of the echinoderm; it also allows the parasites to be grouped in a series illustrating the modifications which the parasites undergo progressively as they bury themselves more deeply into the teguments of the host. Thus the shell, the foot, the head, the gills, the heart and the end of the intestine regress or disappear while the reproductive apparatus is still normally developed. Hermaphroditism is common or, where the sexes are separate, there is a marked sexual dimorphism and the males are always smaller than the females. Dwarf neotenic males occur in many forms, that is, individuals in which sexual maturation has anticipated general growth and differentiation so that larval characters are retained to some extent because the precocious development of the gonad inhibits normal development of the individual.

In this way regressive development in parasitic molluscs leaves only those organs that are essential to this very specialised way of life. However, the regression or disappearance of organs is not the only process involved in adaptation of these parasites; a new organ appears which is progressively developed corresponding to the depth of penetration of the mollusc. It is a structure which arises at the base of the proboscis and ends up by completely enveloping the parasite (figure 2·6). The function of this organ, termed a *pseudopallium*, is perhaps initially to aid the flow of secretion from the pedal gland, thereby aiding the proboscis to penetrate the calcareous plates of the host (figure 2·6). Later, as the pseudopallium comes to envelop the body more and more completely it protects the parasite's tissues from the host. In females eggs accumulate in this cavity, which thus has the additional role of a brood chamber. This secondarily acquired function has become of use in a way already indicated. In this connection it is interesting to note that in *Megadenus holothuricola* the eggs are stuck around the shell of the male and that the pseudopallium of the latter is much more developed than that of the female!

Mucronalia palimpedis, which occurs on the surface of tropical star fishes, has undergone hardly any modifications, except that the proboscis is deeply embedded in host tissues as far as the pseudopallium. The eyes are just visible, there is a small foot and a persistent operculum. This form is permanently attached and does not move about like an ordinary predator. In contrast, in *Megadenus holothuricola* the foot is reduced and the operculum has disappeared; the intestine is short and the stomach forms a single gastro-hepatic cavity with the digestive gland. This parasite lives in the respiratory tree of a holothurian but still retains communication with the exterior. The deeply embedded proboscis pierces the wall of the respiratory tree and penetrates the body cavity of the host. A pseudopallium, better developed in the male than in the female (see above), surrounds the thin shell.

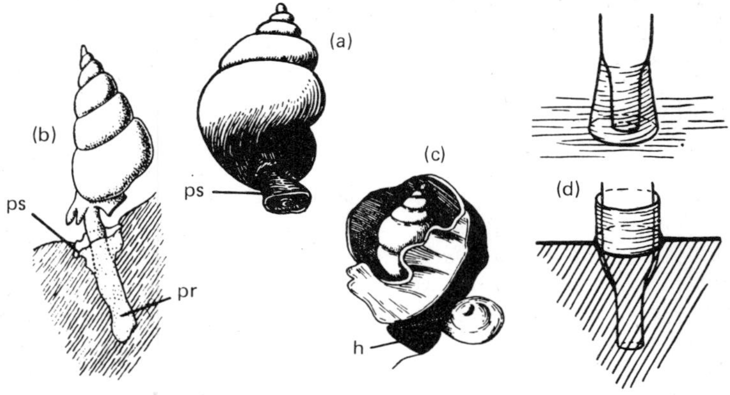

Stilifer linckiae represents a more advanced stage than the two forms described so far because it buries itself so deeply in the body wall of a star fish that only the apex of the shell remains visible at the centre of a small round hole. The proboscis traverses the outer body layers and presses against the peritoneum of the host but does not penetrate it. The body of the mollusc is completely covered by the pseudopallium, which opens by an orifice at the body surface of the star fish. Piston-like movements of the proboscis circulate the sea water in the pseudopallial cavity, facilitating the respiratory exchanges of the parasite. This has a reduced foot and lacks an operculum but the other organs have retained their normal structure. Males are not known and it seems likely that these parasites are protandrous hermaphrodites, that is, forms in which the male reproductive system matures before the female system. In spite of being deeply embedded in its host, this form is relatively little modified by parasitism apart from the considerable development of the pseudopallium. Unfortunately neither the eggs nor larvae are known.

Gasterosiphon deimatis, which lives in the body cavity of a tropical holothurian but has a pseudopallial cavity opening on to the ventral surface of the host, has reached the ultimate stage for an ectoparasite (figure 2·8). The extremely long proboscis is attached within a host blood vessel. The foot is rudimentary, there are neither eyes nor tentacles, the heart and gills have disappeared and

2·6 Development of the pseudopallium. (**a**) *Mucronalia mittrei*, in which the retracted proboscis allows the pseudopallium (ps) to be seen; (**b**) *Mucronalia palmipedis* with the proboscis (pr) buried in the host and the pseudopallium turned back; (**c**) *Megadena holothuricola* in the respiratory tree of its host (h), the pseudopallium forming a brood chamber; and (**d**) a theoretical explanation of the role of the pseudopallium.

the proboscis opens into a gastro-hepatic cavity, but there is no true intestine and no anus. The shell has disappeared but the visceral mass retains its spiral arrangement; the parasite is entirely enclosed within the pseudopallium in which eggs accumulate and this is extended by a short calcareous canal connecting with the surface of the host's skin and attaching the parasite into the body wall. This form is a hermaphrodite so that fertilisation of the eggs is ensured.

An even greater transformation has occurred in *Diacolax cucumariae*, because not only the foot but also most of the visceral mass have disappeared, leaving only a gastro-hepatic diverticulum and the ovary enveloped by the pseudopallium which opens by a short canal. It seems that the sexes are separate, judging from the presence of a receptacle containing sperms, but the existence of dwarf neotenic males cannot be excluded. The eggs hatch during incubation in the pseudopallial cavity and the larvae have a thin shell which is hardly spiralled. This shell is apparently resorbed or cast off – the details are not known – before the larvae are liberated into the sea. They are equipped with two small ciliated lappets, which are, however, not well developed so that they are not able to move far from the host. It is curious to realise that this profoundly modified mollusc is not an endoparasite and that only the proboscis penetrates the body wall while the rest of the body remains external. It would be interesting to know whether this location of the parasite is normal or whether it is due to the reduced size of the unique holothurian found to harbour this mollusc.

The holothurians are also hosts for members of the genus *Entocolax*, which bear a superficial resemblance to *Diacolax* in the almost total regression of the visceral mass and in the development of a large brood chamber which, in *E. ludwigi*, communicates with the exterior at the position where it is attached into the tegument.

2·7 *Stylifer linckiae.* (**a**) The mollusc attaching to and burrowing into a starfish; and (**b**) a section through the established mollusc after the shell has been lost. (pr = proboscis; ps = pseudopallium.)

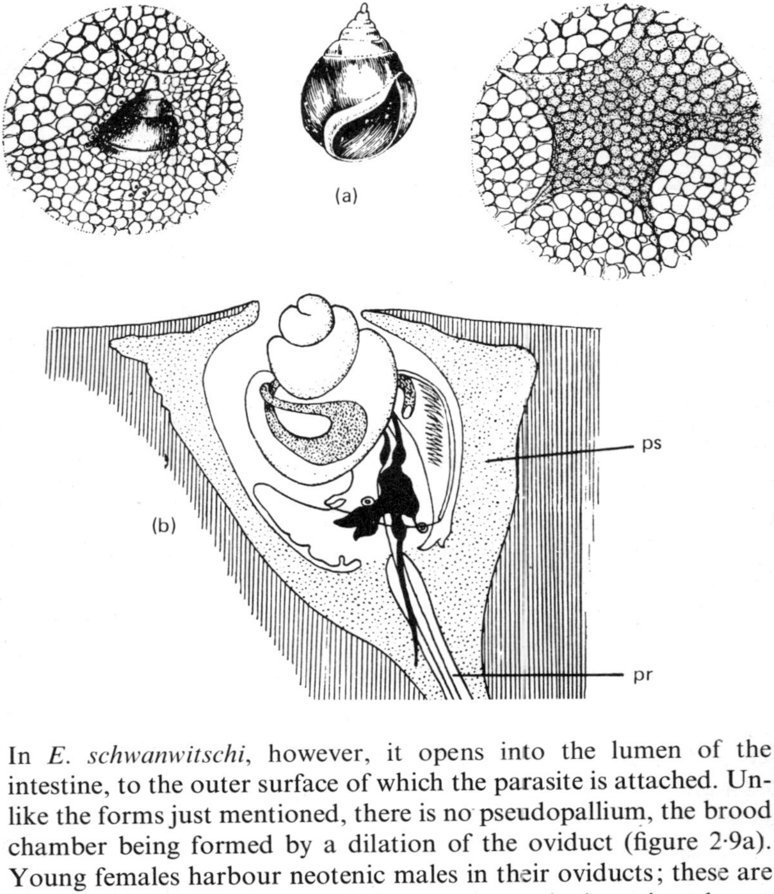

In *E. schwanwitschi*, however, it opens into the lumen of the intestine, to the outer surface of which the parasite is attached. Unlike the forms just mentioned, there is no pseudopallium, the brood chamber being formed by a dilation of the oviduct (figure 2·9a). Young females harbour neotenic males in their oviducts; these are greatly reduced, consisting of little more than a single testis and a vas deferens (figure 2·9e). The egg cocoons are contained in a brood chamber and the larvae have a small, thin shell which shows little trace of spiralling; the foot bears an operculum. In *E. ludwigi* the

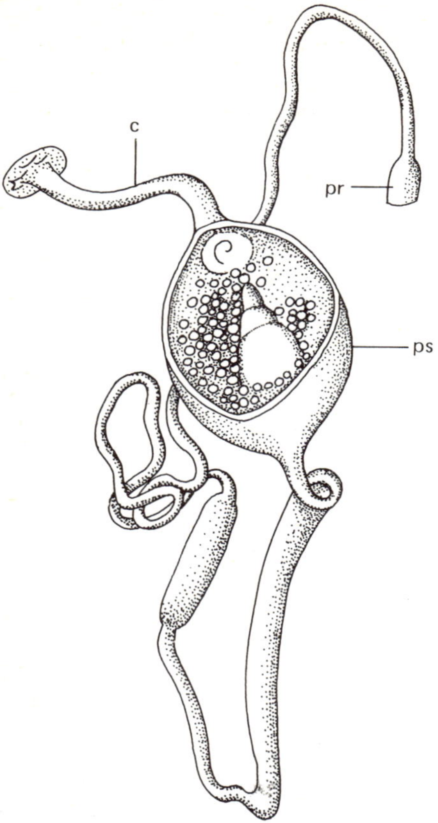

2·8 *Gasterosiphon deimatis* attached inside a holothurian, with the proboscis applied to a superficial vessel. (c = canal connecting the mollusc to the surface of the host; pr = proboscis; ps = pseudopallium.)

larvae are liberated directly into the sea, whereas in *E. schwanwitschi* they leave by the anus of the host and as they are unable to swim actively they crawl in the mud. The larvae of *E. ludwigi* penetrate their host via the outer wall, while those of *E. schwanwitschi* enter via the mouth.

Thus the genus *Entocolax* spans all the intermediate stages between ecto- and endoparasitism; *E. ludwigi* is an example of an ectoparasite and *E. schwanwitschi* is an endoparasite. *Entoconcha mirabilis*, which also lives in the body cavity of a holothurian and which buries its proboscis in a blood vessel, represents an even more advanced stage of endoparasitism because the brood chamber opens directly into the body cavity of the host. Unfortunately little is known about this species, in particular how the brood chamber develops and how the shell-less larvae penetrate the host. This worm-like parasite was actually the first parasitic mollusc discovered, although it was at first confused with the organs of the host; its real nature was not established until some twelve years later.

Parasitic molluscs of the genera *Enteroxenos* and *Thyonicola* (*Parenteroxenos*) represent the final stage in reduction of body parts because only the ovary remains and a brood chamber is formed here by the general body cavity of the parasite; this also encloses a neotenic male which implants itself in the wall, where it was previously mistaken for a male gonad (Lutzen, 1968) (figure 2·9a–d). All the other organs, in particular the digestive system, have disappeared. These molluscs are merely long, semi-transparent tubes about 6–8 cm in length with irregular protuberances, occurring free in the body cavity of the host as in *Enteroxenos* or fixed by one end to the surface of the intestine, as in *Thyonicola*. The parasites are often abundant and holothurians 3·5 cm long have been found to contain thirty to forty specimens. It has been shown experimentally that the larvae gain access to the intestine via the mouth of the holothurian. After casting off the shell the larva digs itself into the wall and as it grows it pushes the host peritoneum before it which comes to invest the parasite very closely. The parasite is nourished by the host and does not need a digestive system of its own. When the parasite is mature and packed with many cocoons it detaches from the wall of the intestine and lives within the body cavity of the holothurian. Expulsion of cocoons containing eggs probably occurs by rupture of the thin distended body wall of the mollusc. However, the cocoons cannot reach the

2·9 Diagrams of endoparasitic molluscs.
(**a**) *Entocolax*; (**b**) *Entoconcha*; (**c**) *Thyonicola*;
(**d**) *Enteroxenos*; (**e**) dwarf neotenic male of *Entocolax*.
(br = brood chamber; h = host; m = neotenic males; o = ova; ov = ovary; od = oviduct; t = testis.)

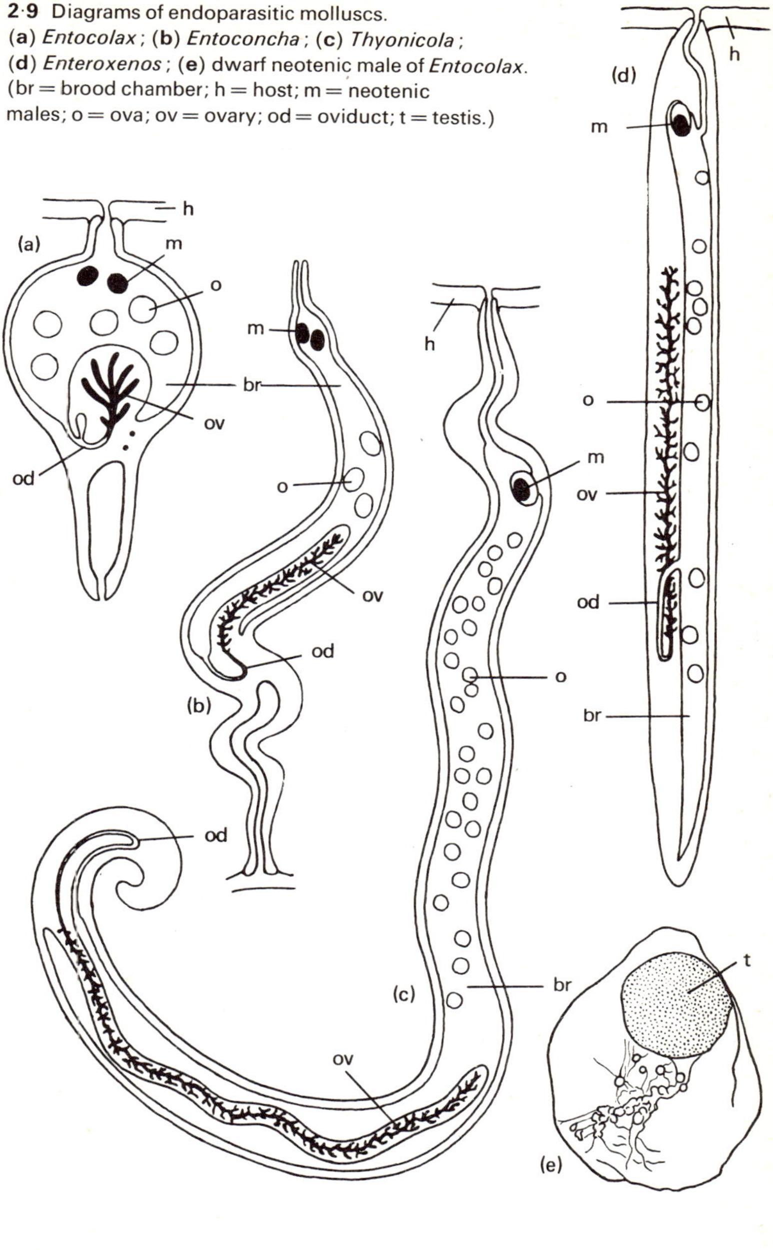

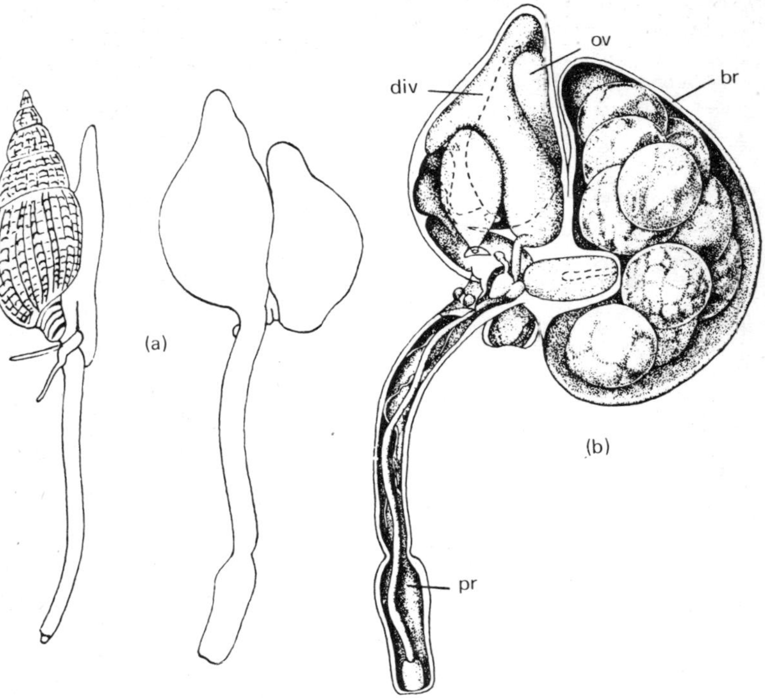

outside until the host casts out its viscera, which is a defence reaction occurring fairly often in holothurians.

Other types of molluscan parasites are not well enough understood to be classed as either ecto- or endoparasites. The form *Paedophoropus* must be mentioned, however, because the brood chamber is formed from part of the foot, a fact which has given this form its name. This form has been observed on only two occasions, once inside the Paulian vesicle and once in the respiratory tree of a holothurian. In both cases the proboscis had penetrated the wall of the organ and was attached to the surface of the intestine. The shell is absent, the oesophagus opens into a hepatic diverticulum, but there is no stomach, intestine or anus. The sexes are separate, with the males about half the size of the females. The visceral mass has lost its spiral structure and the heart and gill have

2·10 *Paedophoropus dicoelobius*. (**a**) Comparison between the parasite and a free-living mollusc (*Nassa*); (**b**) section through a parasitic female. (ov = ovary; pr = proboscis; div = diverticulum of the stomach; br = brood chamber with egg-containing cocoons.)

been lost; on the other hand, the ovary of the female is enormous. The foot, which is also usually reduced in parasitic molluscs, is here hypertrophied, especially in the female, where it forms a brood chamber into which the oviduct opens, and also the pedal gland, which produces a secretion that envelopes the eggs to form the cocoon. The larvae are equipped with a thin shell and a spiralled visceral mass but lack a ciliated velum and consequently have to crawl around in search of a host. Again these larvae cannot be liberated until the host ejects its viscera.

Conclusions: Apart from loss of some organs, the most striking feature about parasitic molluscs is the appearance of a brood chamber which has become secondarily adapted to this new function and which has without doubt played a significant role in the evolution of the parasitic forms. This chamber enables ectoparasites which are embedded quite deeply in their hosts to maintain a respiratory current, but the brood chamber is important mainly because it ensures that only completely formed larvae are liberated. The selective advantage of this organ for the survival of the species is emphasised when it is remembered that this has arisen independently in four distinct lines of parasitic molluscs and has developed from a different organ in each case, namely the pseudopallium in *Megadenus*, *Stilifer* and *Gasterosiphon;* the oviduct in *Entocolax*, *Diacolax* and probably *Entoconcha;* the body cavity of *Enteroxenos* and *Thyonicola* (*Parenteroxenos*)*;* and the foot of *Paedophoropus*. A brood chamber is not present in the free-living prosobranchs or those that are predators, and consequently the secondary function of this organ can be considered as the principal adaptation to parasitism and one which has assured the success of this way of life. The fate of the shell and the organs which have become redundant due to relationships established between the parasite and its host is a consequence of the parasitic existence and emphasises how much the former has become specialised due to restrictions imposed by the host. Finally, the persistence of pedal glands in the young larvae suggests that they might play a part in penetration of the parasite within the host to the situation where development to adults occurs.

Parasitic crustaceans

Most people associate crustaceans with the edible forms – lobsters, crabs, crayfish, shrimps – which belong to the higher crustaceans or malacostracans. They tend to ignore the entomostracan crustaceans which, hardly visible to the naked eye, form an important constituent of the plankton and represent one of the principal food sources for numerous fish and other animals, both marine and fresh-water.

At first sight it is difficult to conceive how parasitic forms which have undergone often enormous morphological transformations could have arisen from these ceaselessly active, free-living forms. Many different kinds of associations with other animals occur in the Crustacea but few of these relationships are properly understood at present. Nevertheless there are a sufficient number of indisputably parasitic forms to allow morphological transformations to be traced. A feature of particular interest is that the larval development of the parasitic forms hardly differs from that of the free-living types, implying that adaptive modifications appear only in the pre-adult and adult stages. In a few cases, however, the whole life cycle has become adapted to parasitism and some larval stages are also parasites or at least fixed temporarily to a host by means of specialised structures.

Among the entomostracans, the copepods and the cirripedes are particularly rich in parasitic forms. It is preferable to deal with these two groups separately, as the free-living copepods are

2·11 Diagram showing the pronounced sexual dimorphism in the parasitic copepod *Chondracanthus merlucci*.

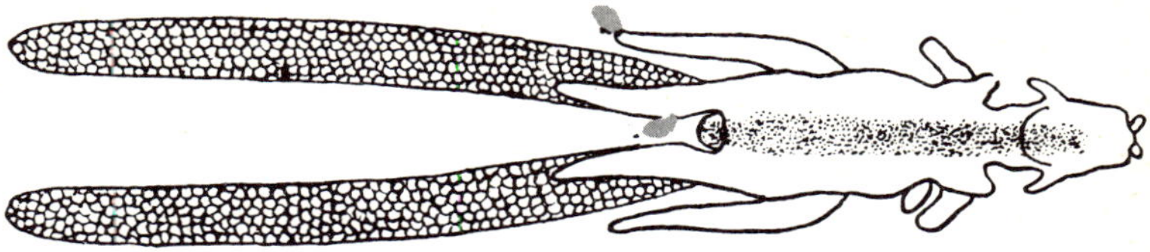

essentially pelagic and have both marine and fresh-water members, while the cirripedes are sedentary, attaching to either animate or inanimate objects, and are exclusively marine.

Parasitic copepods: Parasitic copepods occur in and on a number of aquatic invertebrates, coelenterates, annelids, molluscs, echinoderms, protochordates, on fishes, and even on whales. The parasites range from forms which appear little modified to those which are hardly recognisable due to extreme morphological conformation to parasitic life.

Most parasites have a mouth adapted as an organ of perforation to facilitate fluid feeding. The head appendages are used to establish and maintain contact with the host and are often greatly reduced, being lost in those forms buried deeply in host tissues. A pronounced sexual dimorphism with neotenic males occurs: that is, sexual maturation is precocious during larval development so that growth of the males is inhibited and they remain dwarfs. The difference in size between males and females is often very great, for example 1:12,000 in *Chondracanthus merlucci* (figure 2·11).

Unlike the free-living copepods, the parasitic forms either lay eggs several times during the year or lay eggs continuously. It has been calculated that *Ergasilus sieboldi*, a parasite of fresh-water fishes, lays more than a hundred million eggs a year, in four batches.

The life history of all copepods, whether free-living or parasitic, involves a set number of larval stages between egg and adult (figure 2·12). Six *nauplius* stages and five *copepodid* stages are recognised on the basis of the appearance of different pairs of appendages, so using these distinctions it is possible to identify the time of onset of parasitism. This need not occur simultaneously in

2·12 Larval stages of a free-living copepod.
(**a**) Nauplius; (**b**) copepodid;
(**c**) adult female with egg sacs.

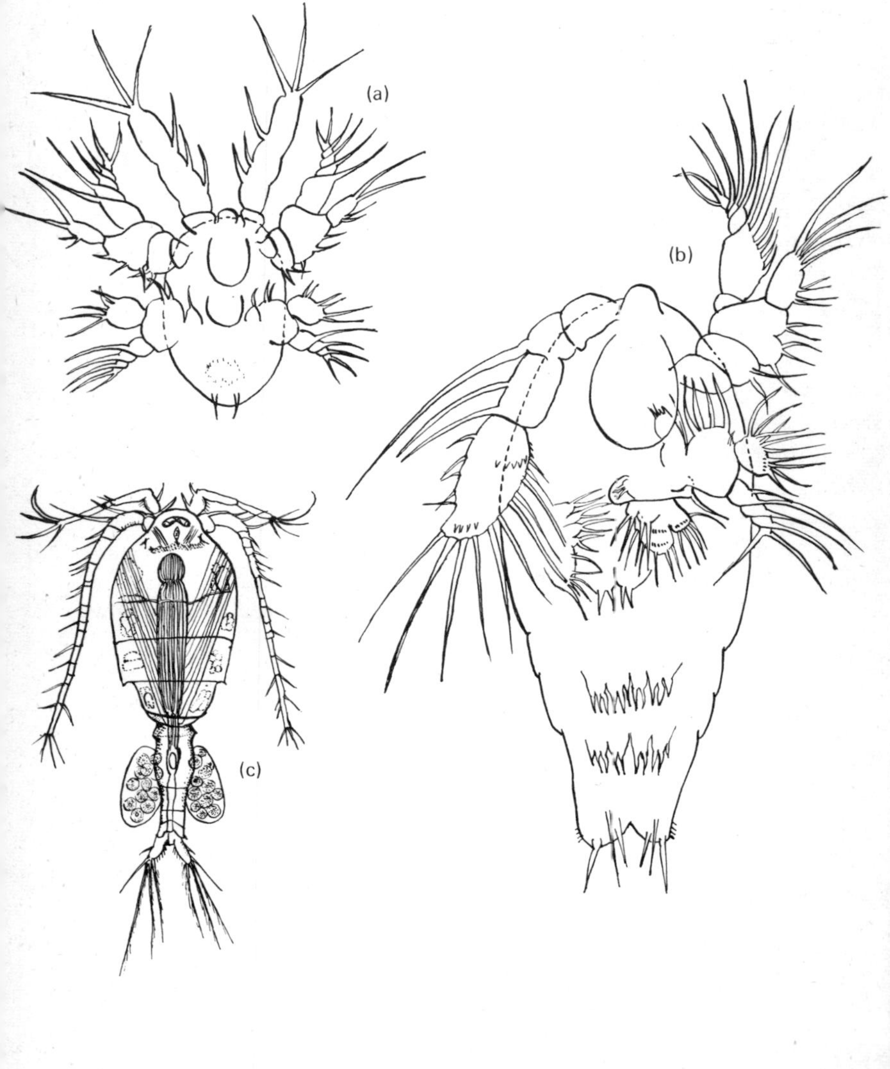

2.13 The parasitic copepod *Caligus rapax*.
(**a**) Nauplius; (**b**) copepodid (chalimus);
(**c**) female adult; (**d**) male adult.

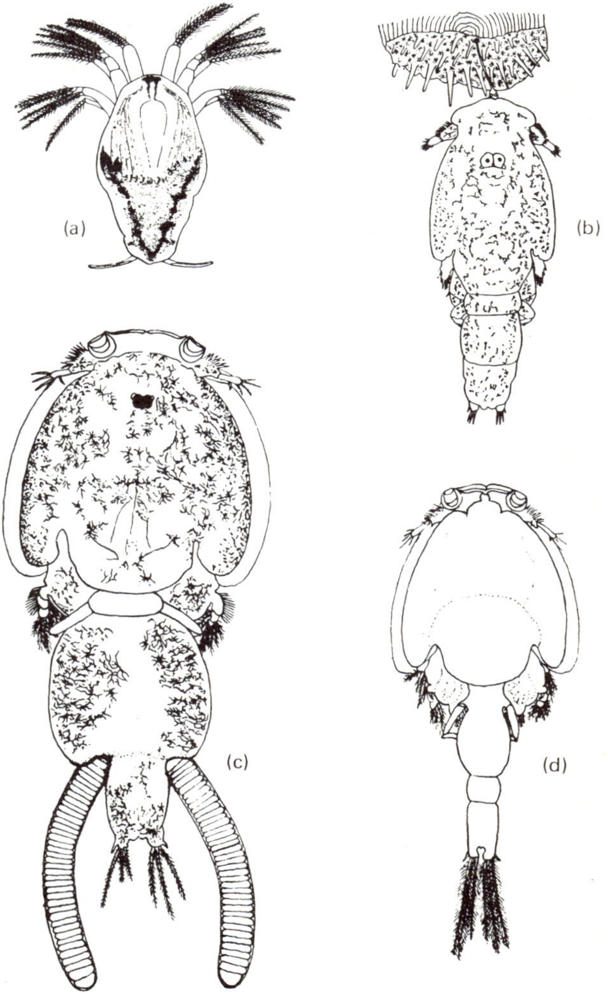

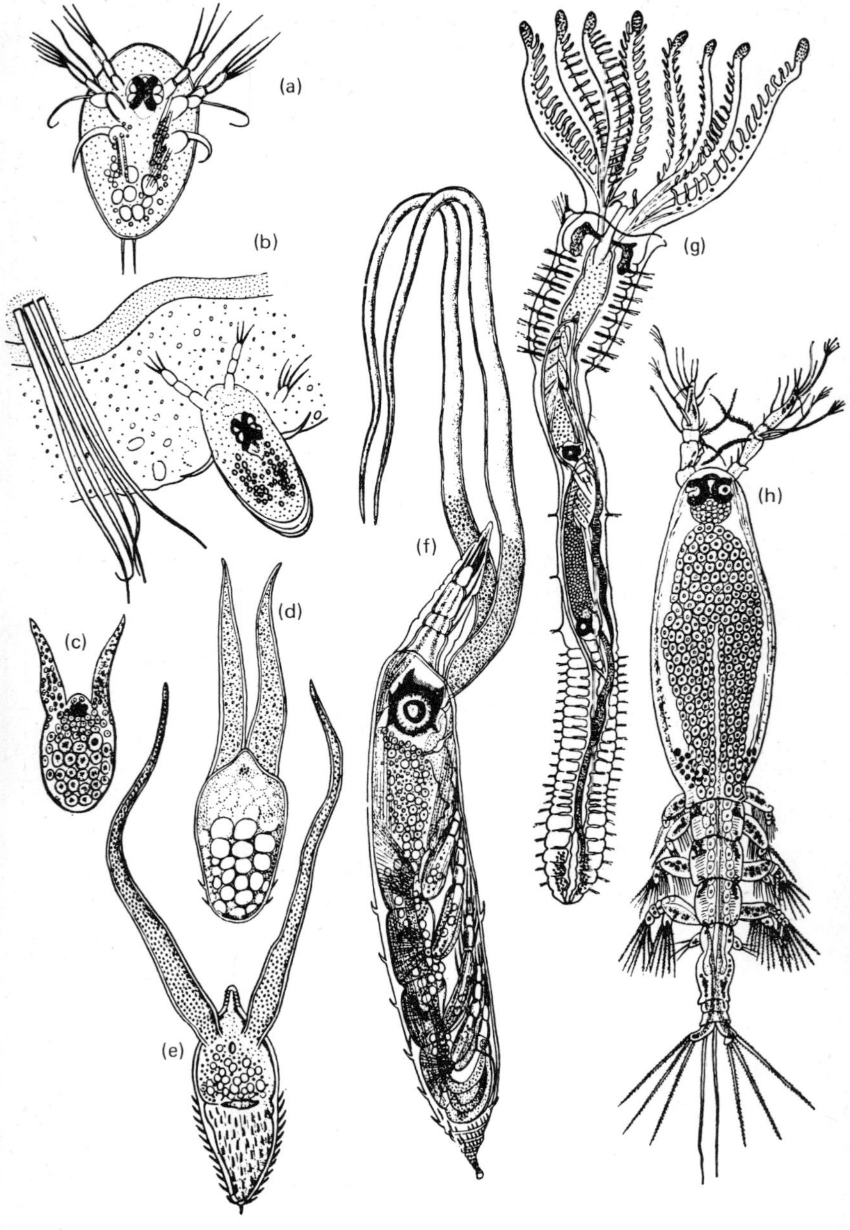
(a)
(b)
(c)
(d)
(e)
(f)
(g)
(h)

2·14 Opposite Life cycle of *Cymbasoma rigidum* (monstrillid). (**a**) Nauplius; (**b**) nauplius in process of penetrating the body wall of the annelid host; (**c – f**) stages in the development of the metanauplius and copepodite showing the development of the two processes used by the parasite to absorb nutrients; (**g**) an annelid harbouring two copepodites in the haemocoel; (**h**) the free-living adult copepod, which lacks a mouth and gut.

2·15 Below (**a**) Male and female of *Linaresia mammillata,* a copepod living inside corals; (**b**) section of a coral showing the fixed females and mobile males.

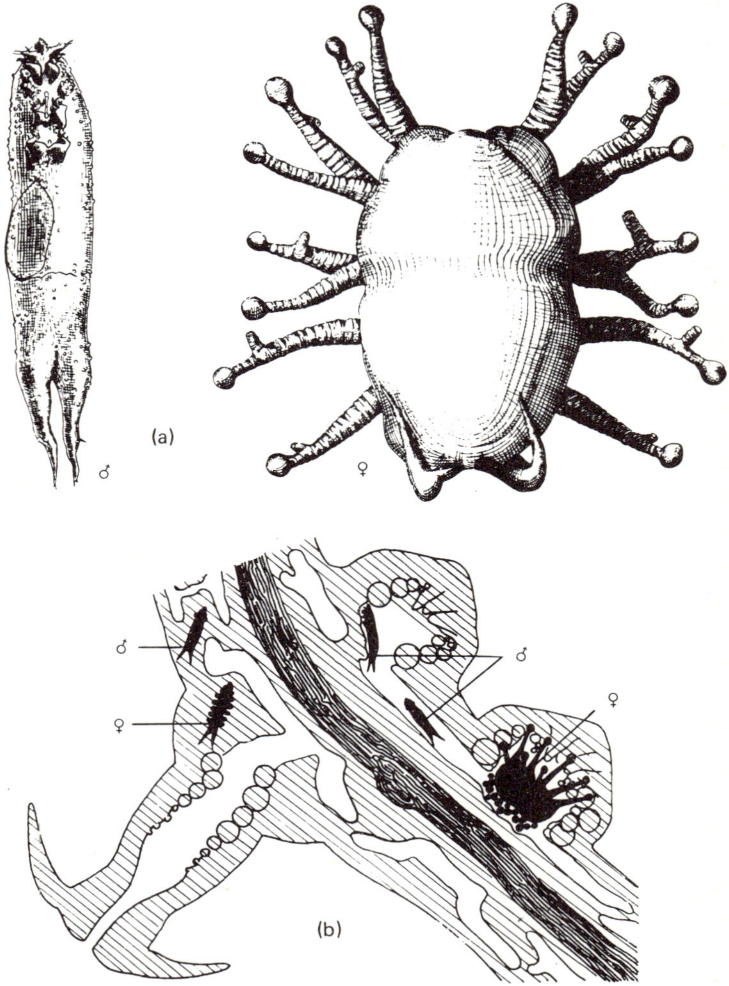

2·16 (a) *Lernaeenicus sprattae* with the cephalo-thoracic region buried in the retinal artery of the host;
(b) *Collipravus parvus* attached within the bulbus arteriosus of its host.

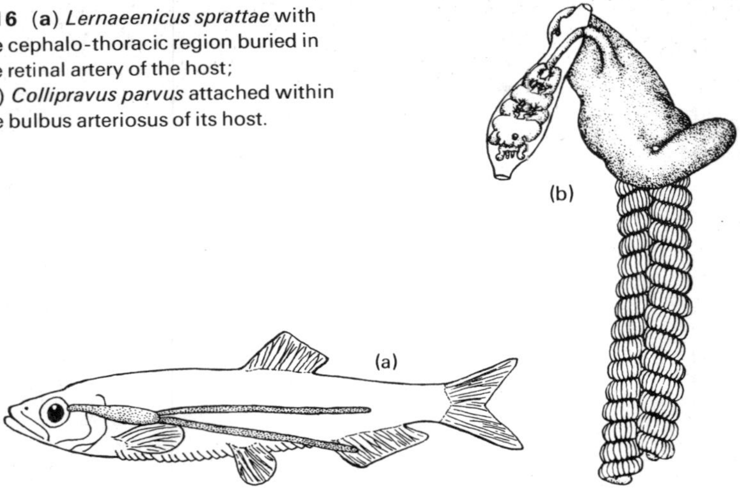

the two sexes of a single species of parasite. For example, in the ergasilids, which resemble free-living forms, only the females parasitise the gills or the walls of the gill chamber of fishes, the males being free-living in the plankton. In the caligids, on the other hand, the adults move on to the surface of the host and can move about freely but the last larval stages and the pre-adults are attached parasites. The head gland of the copepodid secretes a substance which hardens to form a filament which sticks to a scale or to the fins of a fish (*chalimus* larva). Attached in this way the copepodid undergoes several moults. After being fertilised the female detaches itself and continues its existence in contact with the general body surface of the fish. In the chondracanthids the larva attaches to the fish as early as the nauplius stage; the females attach directly to the host but the dwarf males live on the females (figure 2·11).

The monstrillids occur in the plankton and are marine copepods which lack a mouth and gut. They rely on food reserves accumulated during a parasitic larval existence to reproduce before they die. The nauplius burrows through the body wall of a polychaete annelid into a blood vessel, where larval life is completed. During this larval existence there appears a structure which facilitates

2·17 The development of *Lernaea branchialis.* (**a**) Nauplius; (**b**) young female; (**c**) female fertilised immediately after leaving the first host; (**d**) adult male; (**e**) fertilised female attached to the second host; (**f**) adult female, showing the branching head processes that attach the parasite to the gill arch of the host.

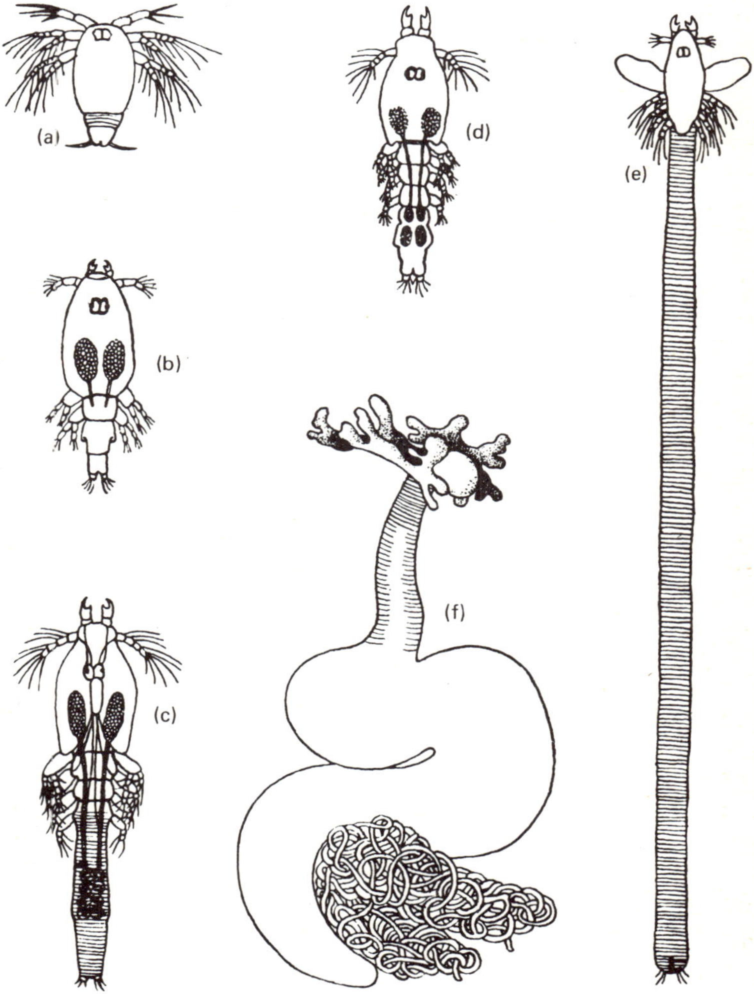

uptake of nutrients but which is rejected when the adult leaves its host (figure 2·14). It is difficult to know whether the absence of the gut here is a cause or a consequence of parasitism. There are, of course, other marine invertebrates, the pogonophorans for instance, which are not parasites but which lack a gut. The adult female of *Linaresia mammillata*, a sedentary copepod living in coral polyps, lacks a gut but has lateral projections which absorb nutrients. In contrast the male of this species has a functional mouth and intestine and moves about freely in the enteron (Bouligand, 1960) (figure 2·15).

The larval development of lernaeids commences in the egg and this hatches to release a late nauplius which attaches to a fish and develops into an adult. Only the males, however, will remain on this first host. The females, once fertilised, detach themselves in order to attach definitively to a new host belonging to a different species of fish. Here they undergo a metamorphosis which leaves them totally unlike copepods except for the persistence of egg sacs. Deeply embedded in the tissues of their hosts the lernaeids often grow to a considerable size (15 to 25 cm). Aristotle and Pliny observed these parasites in the Mediterranean on tunny fish and on the sword fish and deduced that they irritated the fish and made them jump out of the water!

Female lernaeids burrow into their hosts as far as the blood vessels or even into the heart; each species seems to have a specific site: *Cardiodectes* and *Collipravus* in the ventral aorta; *Lernaeocera* in the auricle of the heart; *Lernaeenicus sprattae* reaches the retinal artery by burrowing behind the eye (figure 2·16); and *Lernaea branchialis* attaches to the gills of cod and its metamorphosis involves not only the head but also the thorax and abdomen (figure 2·17). Lateral outgrowths of the thorax penetrate host tissues and progressively ramify. As soon as the female is anchored in this way, the abdomen twists up on itself, increases in volume and

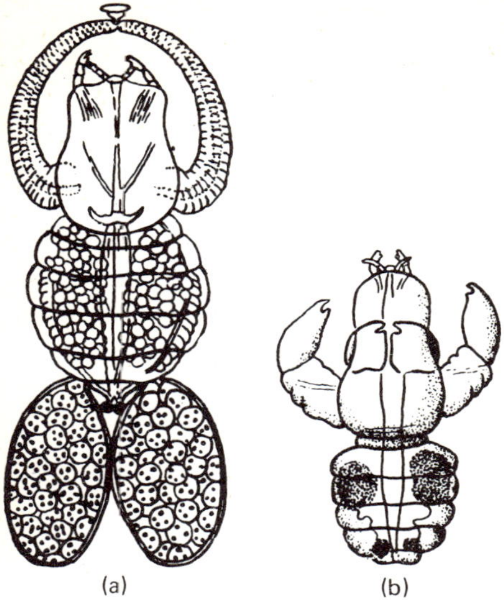

2·18 *Achtheres percarum.*
(**a**) Female with the attachment 'button' which fixes it into the skin of the host;
(**b**) male, greatly enlarged.

shortens. The egg sacs roll up into tight spirals and the parasite will not, henceforth, leave its host.

In lernaeopodids, larval development proceeds as far as the first copepodid stage in the egg so that the cephalic gland has already produced a filament with a round button-like end as in the caligids (see p. 46). The copepodid tears the skin of the fish, using hook-bearing appendages of the head, and inserts the terminal button; the skin then heals over this so that the parasite is attached. At the following moult all the locomotory appendages are lost. Adult males detach themselves by breaking the filament; they then copulate and die. The females, on the other hand, remain attached by the filament; they grip the terminal button with the second pair of maxillae which fuse to it, anchoring the parasite in a definitive way to the gills of the fish (figure 2·18).

Xenocoeloma is a parasite of annelids which lives in the coelom of its host and pushes out the wall of the coelom as it grows. Thus it appears as a cylindrical outgrowth perpendicular to the longitudinal axis of the host, bearing egg sacs at one end (figure 2·19). There is no trace of a gut but it seems likely that the parasite is nourished by soluble substances which diffuse from the host

2·19 Diagram and photograph of *Xenocoeloma brumpti* on the annelid *Polycirrus arenivorus.* The skin of the annelid completely covers the copepod except for the egg sacs, which are free.

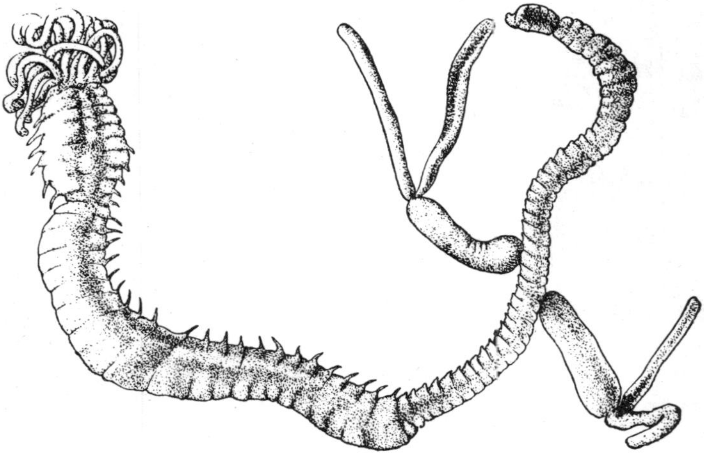

intestine into an axial cavity, around which the reproductive organs are grouped. The parasite literally finds itself in the skin of the host which has been closely involved in the growth of the parasite and comes to invest it tightly. With regard to the larval development of *Xenocoeloma*, all that is known is that the egg hatches to produce a nauplius lacking a gut; the stage of development at which the parasitic habit is adopted is not known.

Many species and genera of copepods are known to be associated with corals, molluscs, annelids, echinoderms and protochordates, but it is often difficult to judge whether they are really parasites, or are commensals, or simply predators. One example has already been cited to illustrate the morphological changes undergone by the female. It is tempting to put this down as one of the consequences of parasitism but it would be premature to assume this. All that can be said is that the more sedentary the females become the more they are modified, while the males, which remain free, are

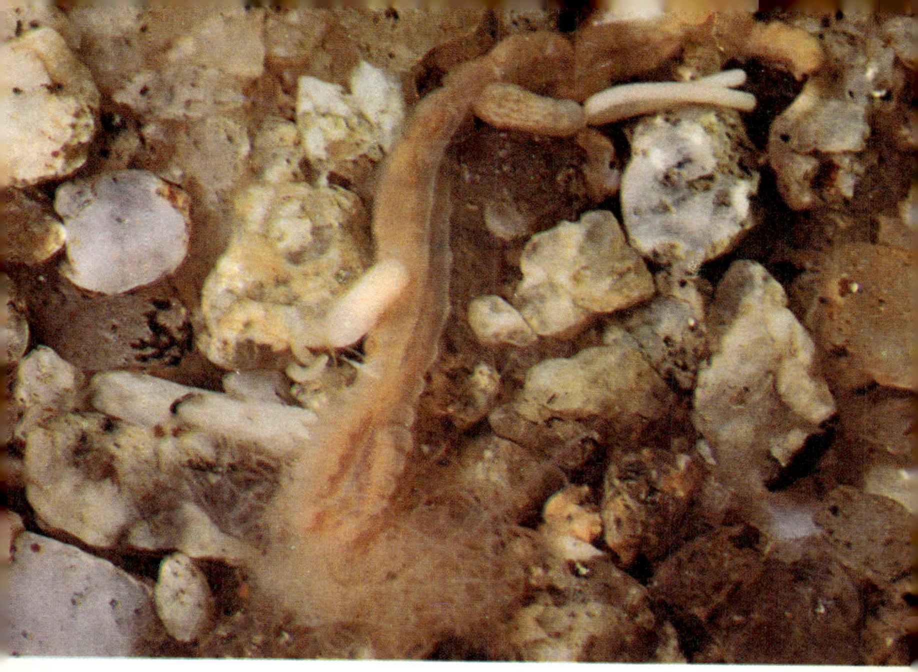

barely affected. A marked sexual dimorphism results which is even more pronounced where the males remain as dwarf neotenic larvae. It is difficult to know whether the parasitic copepods all arose from the same stock; too little is known about their life cycles, and in many instances about their detailed anatomy, to permit a definitive opinion. It is in fact impossible to assess the true relationships of the highly modified adults without this kind of information, especially as most forms are known only through preserved specimens and have been seen only rarely while alive on the host.

Parasitic cirripedes: Walking on the shore at low tide, one often sees on the rocks and other objects small calcareous shells roughening their surfaces. These are barnacles, which resemble the molluscs in having calcareous plates but are really sedentary, permanently attached crustaceans. The nature of the support is immaterial as long as it is firm and barnacles do in fact spread in great numbers

2·20 Below *Anelasma squalicola* attached to the skin of a shark by means of a rootlet system but still retaining a digestive system, albeit reduced.

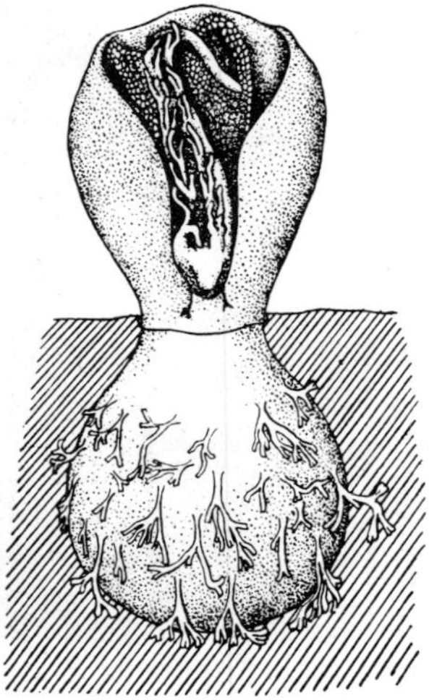

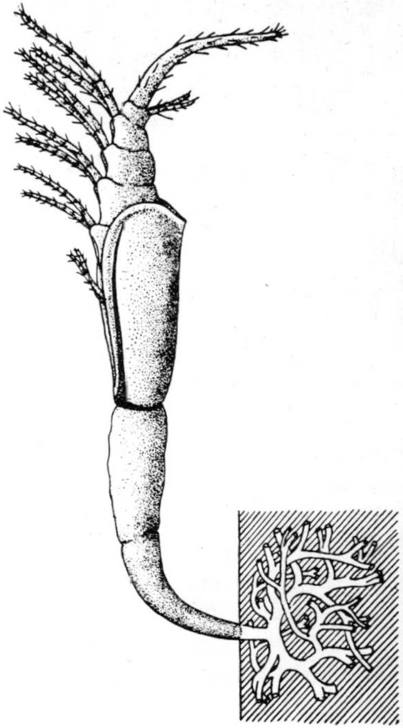

over the submerged hulls of boats, slowing them down and making it necessary from time to time to take the boats into dry dock to scrape them off.

These curious crustaceans belong to the cirripedes, a group related to the copepods. Their development is similar: in both groups a nauplius with lateral horns hatches from the egg. After five moults a stage corresponding to the copepodid is produced but this has the form of an organism enclosed within two thin shells fused along the greater part of their edges. This larva, called a *cypris*, attaches by means of its antennae and metamorphoses into an adult balanid.

At first sight it might seem that the sedentary habit has favoured adaptation of the cirripedes to parasitism. Some species do occur on whales and on the plastron of turtles, but these can hardly be considered to be parasites (figure 1·1).

The small shark *Etmopterus spinax* has a spine in front of the

2·21 Opposite *Rhizolepas annelidicola* attached to its host. The bent part of the body is equipped with a rootlet system by means of which nutrients are absorbed.

fins and in the pocket of skin thus formed at the base of the dorsal fin there occurs a cirripede, *Anelasma squalicola* which buries its posterior end in the tissues while its anterior is visible at the surface. This visible part resembles the head of a normal cirripede because the mouthparts and antennae are hardly reduced. The swollen posterior part of the body, on the other hand, bears numerous ramifying, filiform outgrowths which extend throughout the tissues of the host, causing them to liquefy. Although *Anelasma* also feeds in the manner of free-living cirripedes, this example could be taken to represent an evolutionary stage which has remained at the threshold of total parasitism. This last stage has in fact been attained by *Rhizolepas* which lives on an annelid and has a branching root system which penetrates the host and comes to surround the intestine, thereby anchoring the parasite into the host's body. In the anterior portion of the parasite remaining outside the host, the mouthparts are reduced and the intestine no longer opens to the exterior. This suggests that this truly parasitic cirripede feeds exclusively at the expense of its host.

While it is possible to imagine the kind of transition that has occurred during the evolution of parasitism in some cirripedes, this is not possible for the *rhizocephalans*, which are such profoundly modified parasites as adults that only knowledge of their larvae allows them to be recognised as cirripedes, and this despite the fact that their cypris larva shows parasitic adaptations which are not fully understood.

Although several species of rhizocephalans are known, all of them parasites of decapod crustaceans, comparatively little is known about their life cycles, which makes it somewhat difficult to generalise about this group.

In *Sacculina carcini* the nauplius resembles that of other cirripedes except that it lacks both mouth and intestine. It is free-living and does not feed until the moult that produces the cypris larva.

2·22 Below *Sacculina* (bright yellow) attached to the abdomen of a crab.
2·23 Opposite The life cycle of *Sacculina carcini*. (**a**) Nauplius; (**b**) female cypris larva; (**c**) a larva attached at the base of a spine on the host in the process of casting off its limbs; (**d**) kentrogon larva rejecting the envelope of the cypris; (**e**) kentrogen larva about to insert its dart; (**f**) the female embryonic mass inoculated into the crab.

There are two kinds of cypris, males and females, and only the latter seek out and attach to the base of a bristle on a crab by means of their antennae. At the following moult the locomotory apparatus and accessory muscles are lost, leaving only a mass of cells which give rise to a hollow stylet. The rest of the cypris larva then degenerates, leaving only what is now termed the *kentrogon* larva attached to the crab (figure 2·23d). The stylet pierces the body wall of the crab as far as the body cavity and the cell mass then passes through this into the body of the host so that the larval parasite is actually injected into the crab. This larval mass now proceeds to grow, differentiating into two main parts: a complex root system

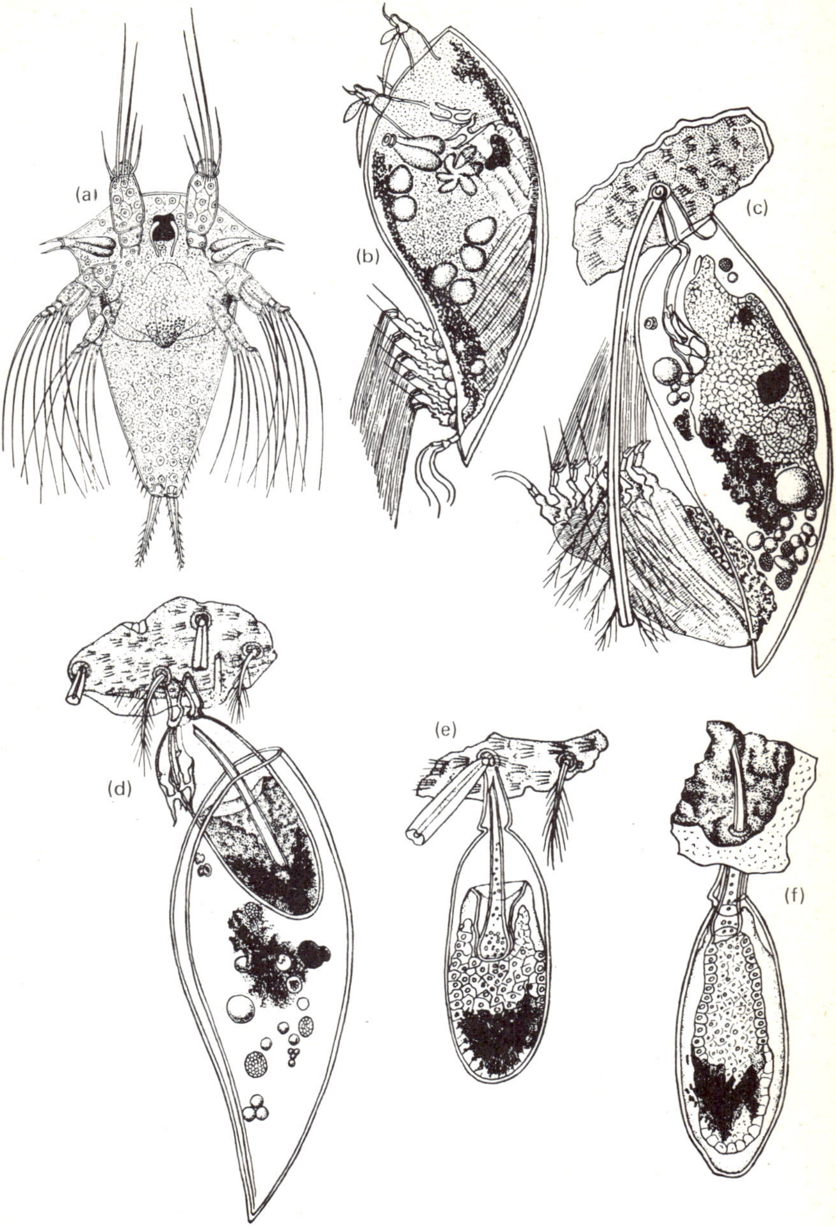
(a)
(b)
(c)
(d)
(e)
(f)

which gradually extends throughout the body of the crab as far as the limbs and serves to absorb food material – this comprises the *internal sacculina* – and a rounded outgrowth which appears after the root system forming the true body of the adult parasite, or *external sacculina*. As this increases in size it becomes applied to and presses upon the ventral abdominal segments of the crab, bringing about the disruption of the ventral body wall muscles and chitin-forming layer of the tegument so that the thin chitin layer eventually gives way, allowing the external *Sacculina* to emerge. This resembles a broad sac enclosing the female reproductive organs and is attached to the internal root system. The enclosed brood chamber communicates with the exterior by a small opening through which the larval cypris males, which are neotenic, enter in order to fertilise the female. It is of interest to note that the neotenic male cypris larvae inject spermatogonia into the brood chamber or into a fold in the wall of this, as in *Thompsonia*. These cells develop rapidly to form spermatozoa in *Sylon*, but in other species the spermatogonia become enclosed in receptacles previously thought to be testes. *Sacculina carcini*, on the crab *Carcinus maenas*, reproduces throughout the year but has a peak nauplius production between the months of August to December. Understandably the activities of the parasite affect the physiological state of the host; for instance, the crab ceases to moult. The fact that a crab already harbours *Sacculina* does not confer immunity against reinfestation. Cases where the crab harbours two, three or even four parasites are not particularly rare. In *Thompsonia* and *Peltogasterella*, parasites of paguran crabs, the external part of the parasite detaches from time to time and is formed anew by the root system.

Parasitic isopods: The parasitic isopods, especially the epicarids, are remarkable amongst crustaceans. Not only are the females, which are ectoparasites on other crustaceans, highly modified, but

2·24 Development of the rootlet system in the internal *Sacculina*.

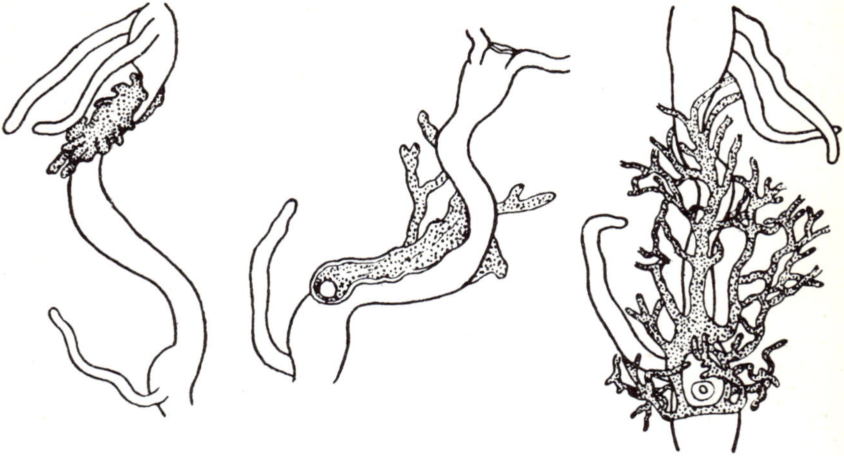

2·25 Below The life cycle of a bopyrinean. (**a**) Microniscus larva which will attach itself to a copepod and develop into (**b**), a cryptoniscus larva which leaves the copepod to attach to the definitive host; (**c**) adult female with dwarf neotenic male. The pereiopods are hypertrophied and form a brood chamber.

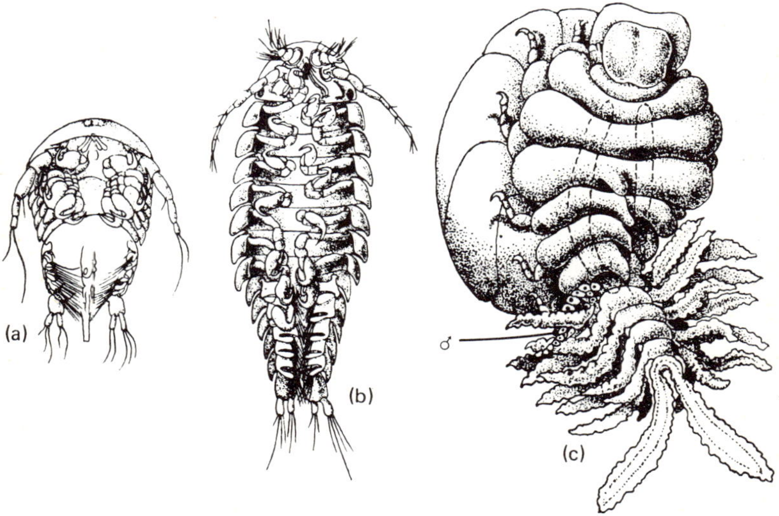

2·26 Life cycle of an entoniscian.
(a) Male, much enlarged; **(b-e)** development of the female with modification of the oostegites to form a brood chamber and transformation of the pleopods into respiratory organs.

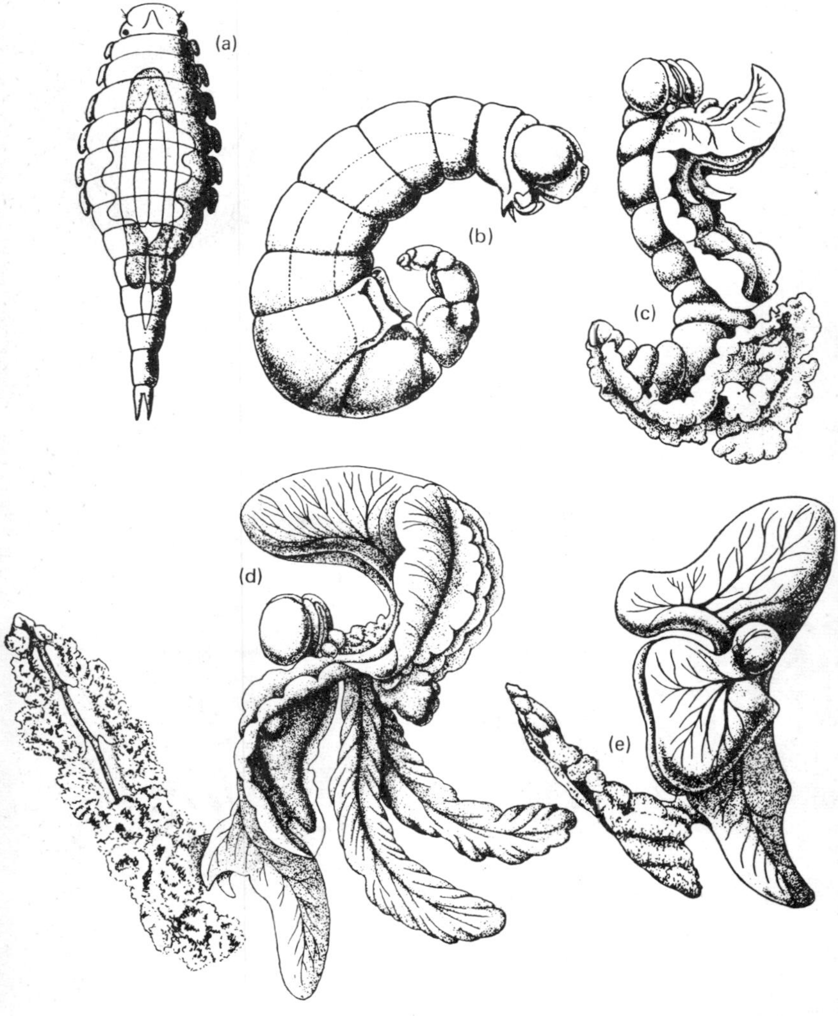

they are parasitised by dwarf neotenic males; furthermore, sexual differentiation is decided by external factors once the parasite has become attached to the host.

In all isopods the female has a brood pouch formed from the ventral plates of the anterior limbs. The eggs are retained here until they hatch. The first larva of the epicarids is termed an *epicaridium* and this swims in the plankton until a copepod is located to which the larva becomes attached. After six moults it transforms successively into a *microniscus* and then a *cryptoniscus*. It was at one time thought that two distinct species of copepod parasites were being dealt with rather than that these were two successive stages in the same life cycle. At this stage the larva leaves the copepod and seeks out a decapod crustacean to which it affixes and develops into an adult. From this point two kinds of life cycle can be distinguished according to whether the larva belongs to the Bopyridae or the Cryptoniscidae.

The Bopyridae are divisible into the Bopyrinae and the Entoniscinae. In the former the larva enters either the branchial cavity or the brood pouch of the host; here it moults and loses the posterior appendages (pleopods). This new stage is named the *bopyridium* and precedes the appearance of sexually mature individuals. In fact the first sexually undifferentiated larva that attaches to the host always becomes a female while other larvae attaching to a host, where a female is already established, always develop into males. If, by chance, several larvae attach to the host all at the same time they all become females but only one develops normally and the others degenerate. It is possible to change the sex of young females by transferring them from their host and introducing them into the presence of an older female. Conversely, males placed on unparasitised hosts will become females. In certain conditions one can even obtain individuals which show characteristics of both sexes (Reverberi & Pitotti, 1942; Reinhard, 1949). Possibly the larvae may be basically hermaphrodite, the appearance of a particular sex being determined by external factors such as food. According to Veillet (1945), *Portunion* has larvae in which the sex is already determined before they penetrate the host. The males are neotenic, retaining some features of the bopyridium and ceasing to grow. While the females increase in size, the males remain dwarf and live as parasites on the female (figure 2·25c). Eventually only a single male retains attachment to the appendages of the female and the latter becomes greatly modified as a result of a considerable development of the brood chamber. Although the bopyrineans still retain some resemblance to highly modified isopods, this is not true of the entoniscians, where the females are not even recognisable as crustaceans (figure 2·26e). These inhabit a cavity lined by a membrane formed by host reaction which opens to the exterior by a small pore through which the epicaridia larvae escape. The parasite feeds on the blood of the host by piercing the host-formed membrane and in this manner obtains oxygen as well as food.

The life cycle of the cryptoniscids is characterised by the absence of the *bopyridium* stage. The *cryptoniscus* larvae attach directly to the host and there they all mature, at first into males, but subsequently they all transform into females. This is due to a kind of protandrous hermaphroditism – that is, in the same individual the male reproductive system is formed first and then, having functioned, disappears and is replaced by the female system which is the definitive phase.

Once attached to the host, the female gorges itself with blood and accumulates food reserves. Then the whole of the digestive system atrophies and disappears together with all the other organs save the ovaries. Signs of external segmentation are rapidly obliterated and the limbs disappear so that all that remains is a sac which pulsates irregularly and is packed with eggs which can only be released by rupture of the body wall and subsequently death of the female.

As can be imagined morphological adaptation of this kind also involves physiological specialisation which manifests itself in the degree of intimacy existing between the parasite and its host. This phenomenon, called *host specificity*, is dealt with in chapter 8.

Conclusions

This chapter reviews the processes by which free-living forms become adapted to parasitism and examples illustrating this have been taken from two major phyla represented by many species distributed throughout the world in both the sea and fresh water.

The molluscs and crustaceans are both ancient groups, as their numerous fossils testify; they have evolved along different lines and have become specialised in different ways. The protelan parasitism of bivalves must certainly have originated as a result of ecological conditions favouring close contact between anodontids and fish. The structure of the glochidia makes it likely that at first they attached to the gills and later, after the appearance of the grappling spines, to the fins. The mutelids have evolved a more complex type of parasitism, for here the larval mantle is adapted to form tentacles. Protelan parasitism has assured the survival of these forms and the disappearance of the fish hosts would doubtless involve the eventual disappearance of the molluscs. It seems likely therefore that protelan parasitism in the anodontids and mutelids was an accidental secondary acquisition which has proved selectively favourable. Although the molluscs can be said to include many predatory forms capable of giving rise to parasitism, this is not true of the crustaceans, so it seems that evolution towards a parasitic way of life must have proceeded differently in these two phyla. The hosts utilised by these two groups are also very different. Parasitic molluscs, live predominantly on echinoderms, a phylum as ancient as themselves, which seems to have evolved but little over millions of years. While the actual nature of the relationship between molluscs and echinoderms and other marine invertebrates is not understood, one would not hesitate to ascribe the term parasite to the rhizocephalans and epicarids living on fish or decapod crustaceans. The larvae of both the molluscs and crustaceans are motile, often very active and capable of seeking out new hosts. Like the crustaceans, the molluscs have followed the route by which predators become parasites but have penetrated more and more deeply into their hosts, in some cases via the skin or via the mouth or anus.

From an ecological point of view, the fact that the locomotory apparatus of the larval endoparasitic molluscs *Enteroxenos* and *Thyonicola* is cast off very early means that they must find a host in the immediate vicinity; larval mortality must be very high because the incidence of parasitism in holothurians is relatively low, despite the enormous number of parasite larvae produced. It is not known whether the parasitic molluscs had a common origin from the prosobranchs but it seems that a primarily predatory existence was replaced by a sedentary habit, implying that at some stage the prey could have been larger than the predator, enabling the latter to feed and reproduce itself without killing the prey and having to find a new 'host'. Isolation on echinoderms must have resulted in specialisation in feeding habits and eventually in morphological changes. The pseudopallium, an organ which protects the tissues of the parasite against destruction by the host, has become adapted as a brood chamber, the selective advantages of which have already been discussed. All other modifications, such as loss of certain organs, are a *consequence* of parasitism rather than an *adaptation* to this way of life. The formation of the brood chamber, whether from the oviduct, the coelom or the foot, is an adaptive process and has favoured survival of parasitism by affording protection to the eggs. So probably the development of the brood chamber is the only true adaptation to parasitism in the molluscs; the four modes of its formation imply the existence of four independently evolving groups. All other modifications, which are often convergent – loss of organs or the shell for instance – are only the secondary consequences of parasitism. The presence of dwarf neotenic males in *Entocolax*, *Thyonicola* and *Enteroxenos*: is discussed later.

With regard to parasitic Crustacea, it seems unlikely that the copepods are monophyletic, that is, that they arose from a common ancestral stock; discussion of this problem is, however, rather premature in view of lack of information on the phylogeny of copepods in general, especially the free-living forms. Phylogenetic links can be traced between forms said to be semiparasitic, hardly differing from free-living forms, and parasitising invertebrates, to highly modified forms parasitising fishes. Some of these 'lines' may show true relationships (see Boquet & Stock, 1963); this group certainly shows an exceptional evolutionary plasticity.

The great fertility of parasitic copepods is probably related to the high protein diet provided by the blood of the host and has certainly contributed to the survival of forms with complicated life cycles, especially those in which the females successively transfer between hosts of different species within the same ecosystem. In general, the greater the hazards encountered in the life cycle the higher is the reproductive output of the females to counter larval loss. This is certainly the result of selective pressures which assure survival of the species and persistence of the parasitic habit.

Adult rhizocephalans bear hardly any resemblance to cirripedes, though their larval development is very similar. In *Sacculina* the female cypris larva attaches by means of its antennae but the most original feature, absent from development of the male cypris, is the kentrogon larva, by means of which the female *Sacculina* is injected into the crab. At least seven genera of rhizocephalans with females parasitising crabs, and being themselves parasitised by neotenic males, are known to date. Some species even show dimorphism as early as the egg and cypris stage, large eggs producing large cypris larvae which are always male and the small eggs giving rise to female cypris which become kentrogon larvae. Female heterogamy is the rule in these cases. It seems that not all parasitic rhizocephalans are females however; a comprehensive account has been given of a parasite of a barnacle which is largely ectoparasitic, does not have a root system, lacks a kentrogon larva, but is hermaphrodite (Boquet-Védrine, 1965).

The bopyrinid epicarids still retain some resemblance to isopods, the main difference being that the females develop a large brood pouch, which reaches extravagant proportions in the entoniscids; the cryptoniscids are no longer recognisable as isopods. The life cycle of epicarids is, however, very different from the direct life cycle of free-living isopods: after hatching the larvae transform by successive moults into males and females. The larval parasitism on copepods and subsequent reattachment to a definitive host with secondary determination of sex is difficult to analyse because of lack of terms of comparison among the isopods. Nevertheless, it may be that the epicarids represent a very old group which have always been parasites of crustaceans. Epigamic determination of sex obviously confers selective advantage on a parasitic mode of life because it assures fertilisation of females attached to a host.

The presence of neotenic dwarf males in both the parasitic molluscs and parasitic crustaceans raises the question of their particular role in adaptation to parasitism. The smaller size of the male alone is not sufficient reason for describing them as dwarf, pygmy or neotenic. The free-living prosobranch mollusc, *Crepidula*, is a protandrous hermaphrodite which is at first male and subsequently female. In this case all the small forms will be males and all the large forms females, but these are only phases of the single, changing individual. Duration of the male phase can, however, be prolonged in several species by the presence of or by contact with female individuals; so far no hormone or diffusible substance produced by the female has been discovered.

Truly neotenic dwarf males occur in the classic example of the annelid *Bonellia*, where the males occur within the uterus of the female, as in certain parasitic molluscs. It has been shown that the proboscis of female *Bonellia* produces a so far uncharacterised substance which brings about neotenous development of the larvae of this form. The most highly modified parasitic molluscs, dealt with above, are bisexual and have dwarf neotenic males which inhabit the brood chamber. This type of parasitism was first noted in molluscs. It is difficult to see how, as in certain copepods, the females have already been fertilised by the time they find the definitive host. The absence of parasitic males other than the neotenics suggests that the female may exert some neotenising action on the larvae or that there may be two kinds of eggs, as in rhizocephalans. Lutzen (1968) has observed a minute ciliated larva lying in the brood chamber of very young *Enteroxenos* in the vicinity of the testis. It seems that these larvae are destined to become neotenic males and have penetrated via the canal still joining the parasite to the gut of the host. This situation has obvious parallels with that in female sacculinids which harbour dwarf males, also with cases where dwarf males occur in the genital ducts of female parasites which are no longer in communication with the exterior, so that eggs and larvae have to be released by rupture of the body wall of the parasite.

As no detailed work has been done on the structure and cytology of the reproductive organs, the role of the neotenic males cannot be definitely established. They occur not only in parasitic molluscs and crustaceans but also in several species of free-living crustaceans; nevertheless, their presence in molluscs in particular must have played a vital role in the fertilisation of females buried deeply in their hosts and no longer in communication with the exterior. No free-living mollusc has so far been found to have neotenic males, so their appearance might well have been connected with survival of this type of parasitism.

The neotenic males of the epicarid crustaceans which are carried between the pleopods of the female ensure the fertilisation of a considerable number of eggs produced as a consequence of a high protein diet. As the host usually supports only a single epicarid the female parasite can exploit its situation without the risk of seriously affecting the host. Male epicarids, like the neotenic male rhizocephalans, do not obtain food from the host, and this too favours development of a stable parasitic relationship and is a secondary adaptation to this way of life.

3 Ectoparasitic insects

This chapter deals only with the obligatory insect parasites of warm-blooded vertebrates represented by the fleas, pupiparan dipterans of the hippoboscid, streblid and nycteribid families, and by the Mallophaga and lice. The first two of these groups have free-living larvae, the third and fourth groups spend their whole existence on the host.

Fleas or siphonapterans

Fleas are insects that undergo complete metamorphosis involving a larval phase, a pupal stage and an adult. Only the latter is parasitic and feeds exclusively on a bird or mammal host, the larva being free-living in the nest or den.

The adult flea is well adapted for moving rapidly between hairs or feathers. At the same time it maintains attachment by means of the comb situated behind the head (figure 3·2) and by the terminal claws and the spines of the legs. The laterally flattened body allows it to slip easily between the bases of hairs or feathers, and its remarkably long third pair of legs enables it to leap in its perilous search of a new host. Fleas often desert their hosts, but it has not been proved that the blood of an accidental host is necessarily suitable for the flea. For example, a rat flea, *Nosopsyllus fasciatus*, will attack man, mice and rabbits but is able to lay eggs only after a meal of rat blood. The ability to use blood from a host other than the original host, even temporarily, does suffice however to prevent the flea from starving while it waits for a host with blood suitable to ensure the development of its eggs. The less a flea is specialised with regard to needing a particular kind of blood to produce viable eggs, the greater is the chance that its leaps will bring it into an environment favourable for egg laying and suitable for the larvae. Eggs laid on the body of the host fall into the nest or into the detritus in the den, and here the larvae are less affected by the fact that the nest belongs to a bird, a squirrel, a rabbit or a mouse, than by the actual conditions such as temperature and humidity associated with this environment. This probably explains why most bird fleas are associated with species nesting close to or on the ground and with those constructing their nests with mud.

The relationship of fleas to their hosts is essentially an ecological one and a population of fleas in a nest abandoned by a bird will soon disperse, though one or two may survive to infest new hosts.

Pupiparan dipterans

This group of parasitic flies is characterised by the quasi-absence of a larva. Eggs hatch in the oviducts of the female, which are supplied with special glands producing a secretion used to nourish the larvae. The latter are deposited at a very late stage of development and pupate almost as soon as they are produced. In general the flies leave the host to produce young so that the insect emerging from the pupa has to find its own host. The hippoboscids are parasites of birds and mammals while the nycteribids and streblids live exclusively on bats.

Numerous morphological adaptations to parasitism can be recognised in this group. It is, however, difficult to decide which features are primitive and which have been secondarily evolved in connection with parasitism. All are blood feeders. They show a marked tendency towards flattening of the body and elongation of the legs. Like fleas, they often have an arrangement of posteriorly directed rigid hairs together with spines and bristles (figure 3·3). The hippoboscids and streblids show all stages in wing reduction, ranging from those with normally developed wings to those lacking them completely. In certain hippoboscids (*Lipoptena*) the wings which are present when the imago emerges from the pupal case

3·1 Below Flea larvae obtained from the burrow of a mole, probably *Ctenophthalmus averenus.* (Natural size 5 mm.)

3·2 Bottom The comb of a flea.

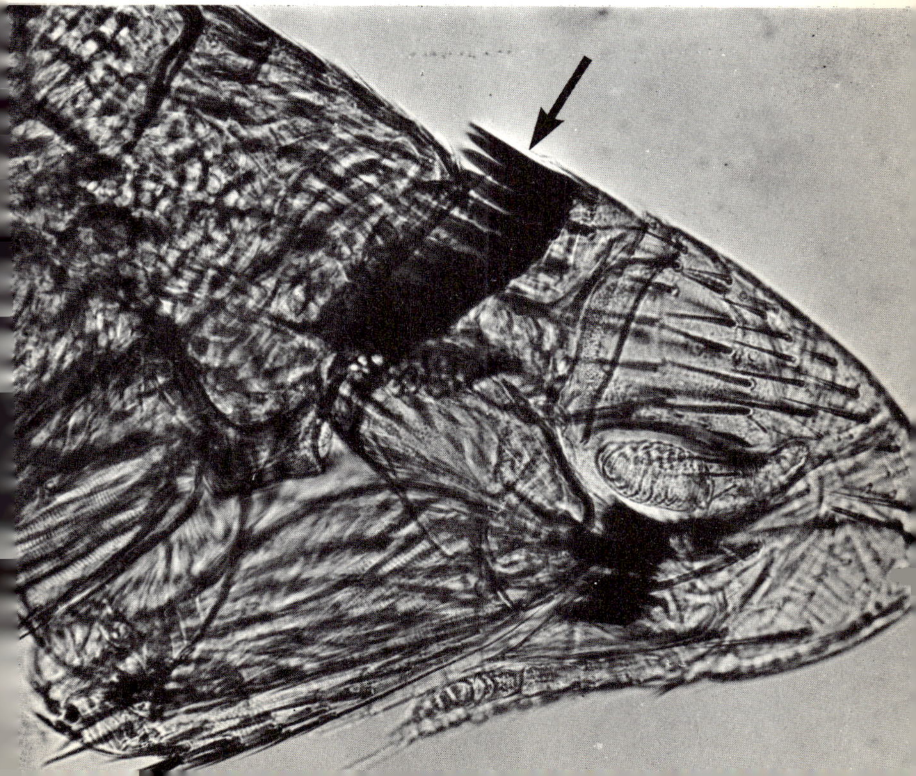

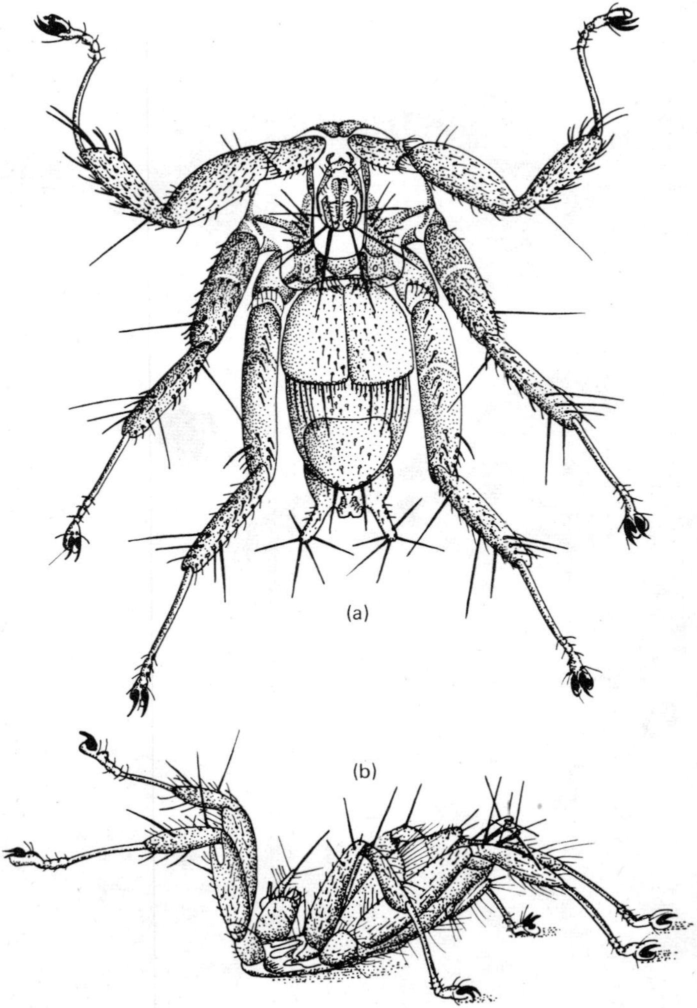

3·3 A nycteribid. (**a**) Dorsal view showing how the wings are completely reduced; (**b**) lateral view of the dorso-ventrally compressed body.

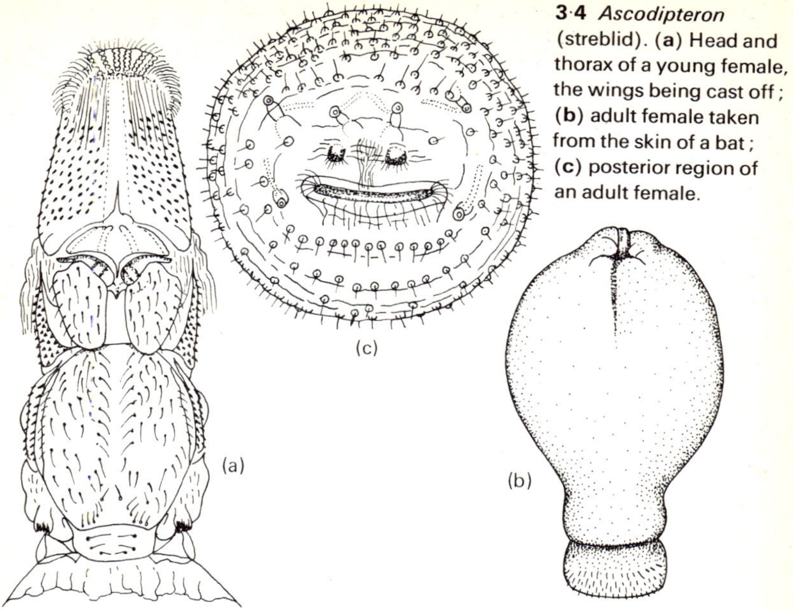

3·4 *Ascodipteron* (streblid). (**a**) Head and thorax of a young female, the wings being cast off; (**b**) adult female taken from the skin of a bat; (**c**) posterior region of an adult female.

are lost when the fly becomes established on its host. The female of the extremely specialised streblid, *Ascodipteron*, loses not only its wings but also the legs by autotomy and buries itself in the host skin. The males remain winged and are free to move over the body of the bat. In contrast, the nycteribids have completely lost their wings. There is also a tendency towards reduction of the eyes. In the hippoboscids atrophy of the eyes parallels loss of the wings, the reduction being most marked in the apterous form *Melophagus*. There is, however, no relationship between loss of the eyes and wing reduction in the streblids, while in contrast the apterous nycteribids have eyes formed of only a few ommatidia. Significantly, in the latter group those with comparatively well developed eyes occur on fruit-eating megachiropterans such as the flying fox which, instead of living in caves, hang from trees during the day.

Although the affinities of the pupiparans and their relationship with other dipterans is not well understood, the tendency towards blood feeding and larviparity has definitely influenced the develop-

3·5 *Myrsidea ishizawai* (Mallophagan), from a passerine bird.

ment of parasitism in this group. It seems likely that the blood feeding habit was acquired independently of larviparity because many dipterans are haematophagous without showing larviparity, while other pupiparans, such as *Glossina*, are blood feeders but are not obligatory parasites. Deposition of larvae in nests or holes or by attaching them to the walls of caves must certainly have encouraged an ecological association between the hosts and their parasites; the evolution of the latter then expressed itself in the secondary acquisition of combs, hairs and spines and finally by atrophy and loss of the wings and reduction of the eyes.

Mallophagans and lice (phthirapterans)

Modern classifications unite the mallophagans or chewing lice and the sucking lice or anopleurans in one group, the phthirapterans. There are, in fact, mallophagans which have elongated, penetrating mouthparts intermediate in type between those of the two traditional groups. Mallophagans occur on birds and mammals, while the sucking lice are exclusively parasites of mammals. The food of mallophagans consists essentially of fragments of keratin derived from hair along with epidermal debris and blood. The sucking lice are essentially blood feeders. At no time during their existence do the phthirapterans have wings. They develop without metamorphosis. The eggs are stuck to the hairs or feathers of the host and give rise to a small louse which grows without leaving the host. Thus invasion of the host is assured and growth of the parasite population depends on external factors, such as temperature, as well as internal factors connected with the metabolism of the host. The relatively small number of eggs produced is compensated for by the fact that the whole life cycle is completed on a single host, which of course reduces the risks involved. The Mallophaga living on the head and neck of birds are short, rounded and only slightly

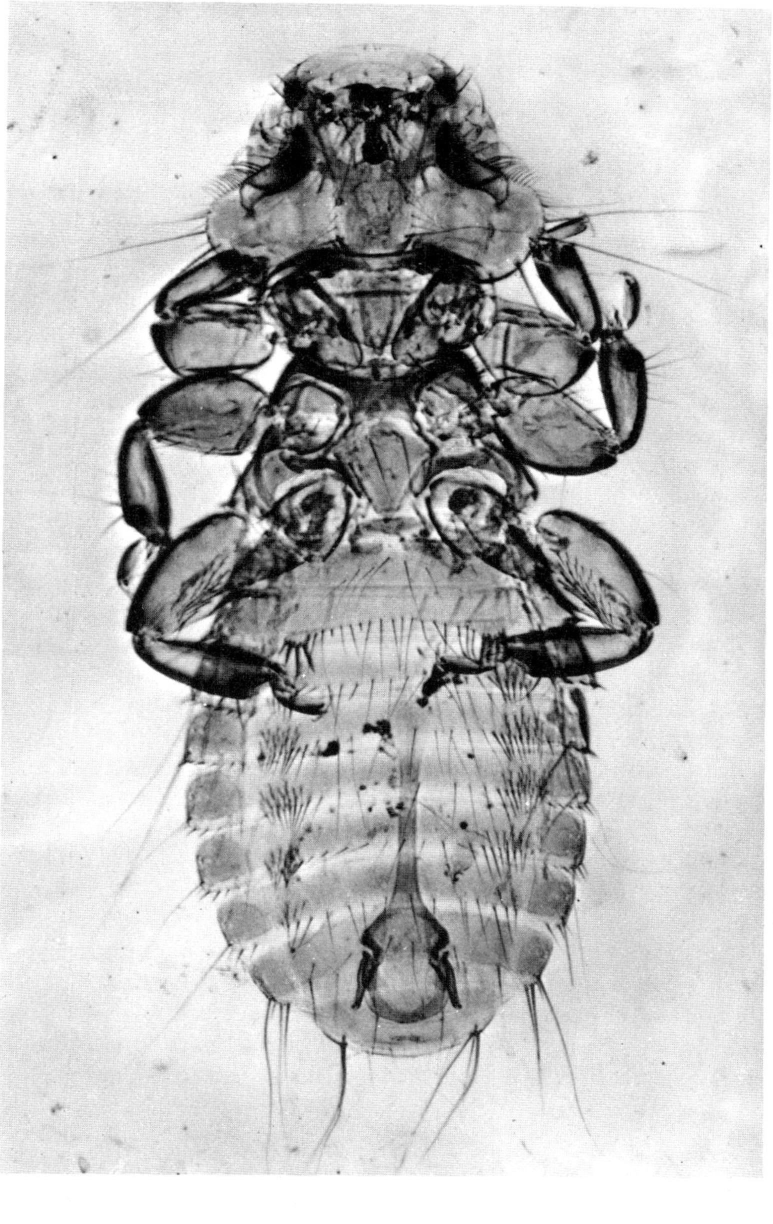

flattened dorso-ventrally, with a large head and powerful mandibles. These forms are adapted to live in between the short, thin feathers in this region of the host and because they are inaccessible during preening they would gain no advantage from being flattened. Forms living among the much larger feathers of the back and wings have flat, elongated bodies allowing them to slip rapidly between the feathers, so escaping being caught by the host's bill. In the first case the eggs are cemented to the short head and neck feathers and in the second case they are placed between the barbs of the large feathers, which affords a high degree of protection. It must be pointed out that not all Mallophaga occupy secure retreats on the host. Many birds have a kind of comblike projection beneath the median toe which is used to scratch the head. The efficacy of this organ is such that hosts possessing it never harbour mallophagans on their heads. A bird prevented from preening effectively by a deformed or damaged beak is much more highly parasitised than a normal bird. Use of naturally occurring insecticides also enables the host to control, to some extent, its mallophagan population. Dust baths are one example, and also formic acid, which the bird obtains by placing itself over an ant hill and allowing itself to become covered with ants which, it seems, attack the mallophagans. It has been observed that birds sometimes rub their plumage with dead ants which they hold in their bills. There seems to be a relationship between the colour of mallophagans and the feathers they inhabit, which is doubtless the result of a natural selection favouring survival of the best camouflaged parasites. Dark mallophagans live in the darkest plumage and the palest forms in white plumage. The swan, for example, bears very pale coloured lice under its wings whilst those on the neck are almost black, protected not by their colour but because they are out of range of the beak.

Selective mechanisms tending to isolate parasite populations in ecological microniches operate in mallophagans and guarantee a succession of generations on the same host. The host at the same time uses innate means or develops secondary methods to control the growth of the mallophagan population.

Infestation of new hosts occurs for the most part in the nest and if a host dies its mallophagan population will perish with it unless transmission to another host of the same species is possible. Accidental transfers by contact or by intervention of a hippoboscid (figure 3·6) to which mallophagans attach themselves are not always successful. Interestingly enough, the young cuckoo never acquires the mallophagans of its adoptive parents but has its own species, which are probably acquired in its winter terrain when in contact with adult birds.

All these isolating mechanisms have led to speciation of the mallophagans, and also to development of a pronounced degree of host specificity, a phenomenon which will be discussed in chapter 8.

Anopleurans are found only on mammals, and their absence from birds may be due to the high avian body temperature of 40 to 44°C. They feed exclusively on blood, using their buccal stylets to pierce the skin. Like most blood-feeding insects, the anopleurans have symbiotic yeasts and bacteria which provide essential substances absent from the specialised blood diet. It has been found possible to maintain human fleas previously deprived of their symbionts on haemolysed blood supplemented with B vitamins. This provides a method by which the functional significance of various vitamins can be assessed. The absence of lactoflavin (B_2), pyridoxin (B_6) and folic acid retards larval growth up to the third instar, and from then on development can be halted by the absence of pantothenic acid and nicotinamide (Puchta, 1955).

It is to be hoped that this line of investigation will be developed, as it could significantly advance our understanding of the physiological relationships between lice and their hosts. Little is known about the vitamin content of mammalian blood. Human blood is, however, known to lack four B-group vitamins which are supplied to the lice by their symbionts.

Lice are not able to move easily on hairless skin and cannot attach their eggs to it. This explains their absence from animals such as the rhinoceros, hippopotamus, whale and pangolin. In fact, among the carnivores only the Canidae are parasitised and the pinnipedes are the only marine mammals carrying lice, these being specialised to live in the fur of sea lions and in the nares of seals. The body louse of man arose from the head louse, but this new habitat, unique for the anopleurans, was occupied only after man started to wear clothes, even where these consisted only of an ornament strung around the body on which eggs could be deposited! Peoples without either ornaments or clothes are also without body lice, although populations may occur on the head.

Like the mallophagans, anopleurans have a particularly intimate relationship with their hosts, and lice can be successfully exchanged only between hosts of the same species.

The ectoparasitic insects, as defined here, are far from forming a homogenous group. They fall into two biological categories, because the pupiparans and fleas develop with metamorphosis and have a free-living larval stage while the phthirapterans have no metamorphosis and all stages from egg to adult live permanently on the host. The absence of wings from fleas limits to a certain extent the choice of hosts and leads to specialisation as a result of their isolation on groups of hosts having similar ecologies. There is reason to believe that the composition of the blood taken in by fleas is important and that only the blood of particular hosts will allow the production and laying of viable eggs. The flea can, however, survive but not reproduce on blood meals from foreign hosts and it is because of this that rat fleas can transmit to man the plague bacillus, originally an epidemic disease of rodents.

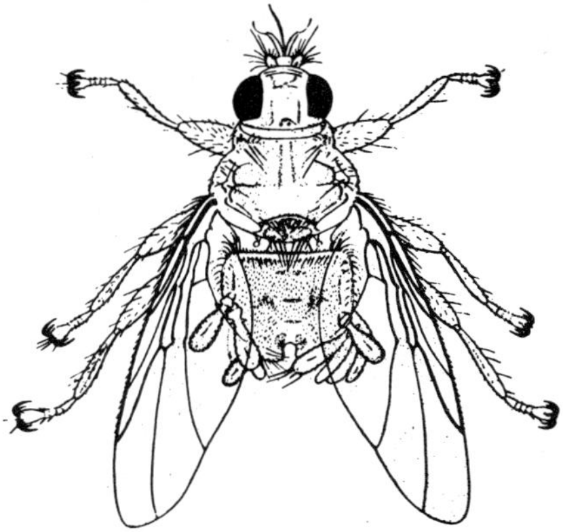

3·6 Right Mallophagans attached to the abdomen of an ornithomyian (pupiparan). In this manner they can be transported to another host.

3·7 Below Anterior region of a body louse.

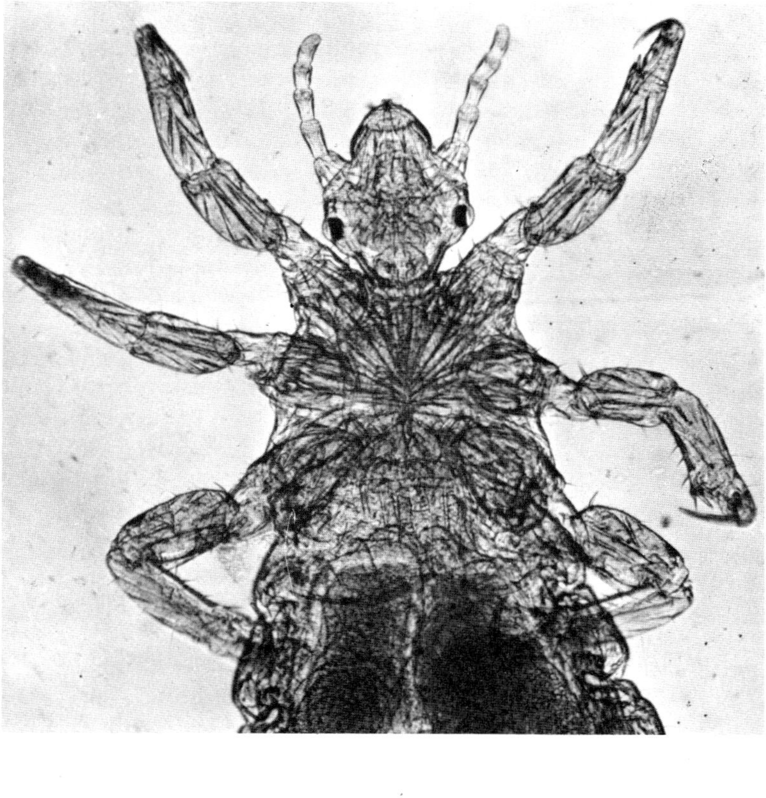

The pupiparans would at first sight seem to offer several examples of adaptation to parasitism, such as reduction of the eyes, progressive loss of the wings, the bringing up of the legs in close association with the thorax and flattening of the extremely hairy body. But it is difficult to evaluate the selective advantage these characters might confer in the development of a more strictly parasitic relationship, governed above all by the blood-feeding habit. The glossinids (tsetse fly), for example, are also haematophagous pupiparans but do not show any of the morphological characters mentioned above. They leave their victims as soon as they are gorged with blood and later attack another, rather like a predator. The female of *Carnus haemapterus*, which belongs to a completely different family of dipterans, has a flattened head and thorax and rejects its wings as soon as it is established on its bird host, where it is an obligatory parasite. Hippoboscids belonging to the genera *Crataerina*, *Mycophthria* and *Stenepteryx* have reduced, stumplike wings and spiny feet ending in claws. This is a special adaptation to life on rapidly flying swallow and martin hosts, which are airborne for long periods and where possession of wings by the parasite might result in them being torn from the host's body. The genus *Ornithomyia*, which is specific to swallows has, however, kept its wings! It is now considered that the glossinids and hippoboscids are derived from a common ancestral group but that they subsequently evolved in different directions. On the other hand, the nycteribids and streblids, associated with bats, which lack wings and eyes, are derived from another branch of guanophile dipterans.

As for the phthirapterans, which are derived from the free-living, winged, algal and fungus-feeding psocopterans, the absence of metamorphosis and the fact that the whole developmental cycle is accomplished on one host has 'predestined' them to a high degree of morphological and physiological specialisation, the consequences of which are studied in chapter 8.

4 Round worms or nematodes

Among nematodes as a whole there are very many which are not parasites and live in many diverse kinds of environment in all latitudes. They occur in the soil, in both fresh water and marine conditions, in hot mineral springs, and have even been found in icy polar regions. Some of them attack plants, causing considerable damage to crops, others associate with insects, which they often kill. Vertebrates have been acquired as hosts on several occasions, yet it is not possible to trace anatomical modifications that can be related to a parasitic way of life. All species, whether free-living or parasitic, have the same basic structure; the body is filiform and elongate and is invested by a longitudinal muscle layer and bounded externally by a cuticle. The mouth is terminal and the anus is at the opposite end of the body. Food material, which may be largely liquid, is pumped into the straight intestine by means of a muscular pharynx. The sexes are usually separate but some forms are hermaphrodite or parthenogenetic.

The free-living nematodes are usually microbiverous, living on decomposing organic material or on bacteria and other micro-organisms; the great diversity of their habitats and their capacity to adapt to widely different temperature conditions suggests how they have been able to exploit the many ecological niches offered by a living organism and become successful parasites. The nematodes are clearly preadapted to parasitism, since larval development is the same in the free-living and the parasitic forms. Free-living nematodes are easily cultured from the egg stage, using suitable media. They develop through five stages separated by four moults, the fifth stage transforming into the adult worm. This can be shown schematically as follows:

$$\text{egg } L_1 + M_1 \rightarrow L_2 + M_2 \rightarrow L_3 + M_3 \rightarrow L_4 + M_4 \rightarrow L_5 = \text{adult},$$

where L_{1-5} are the larval stages and M_{1-4} the moults.

Examination of cultured worms shows that the different larval

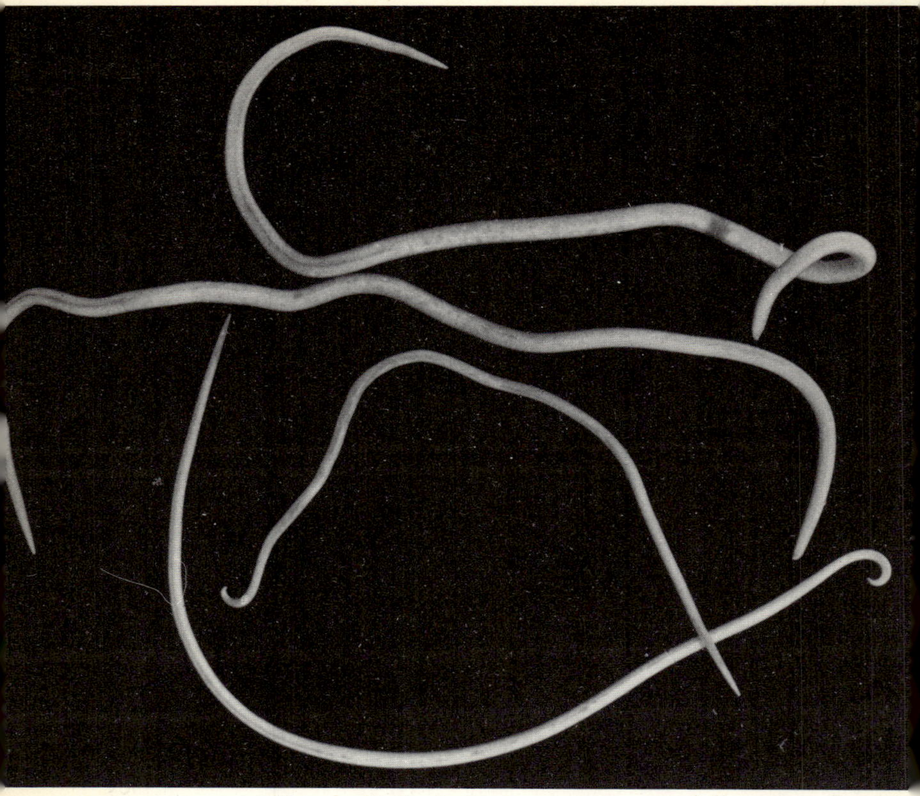

stages have different biological characters, especially stage L_3 which is still sheathed in the uncast cuticle from moult M_2 and is extremely resistant to external conditions, especially desiccation. If the culture were to dry up only the third-stage larvae would survive and these would be able to exist in a desiccated condition for several months or even years in a state of cryptobiosis. The term cryptobiosis is used to mean the ability of organisms to survive long periods of desiccation and become active again in the presence of water. Consequently, moistening the dry and inert larvae with this medium would allow them to become active again and to complete their development. In natural conditions this specialised larval stage aids survival and even dissemination of these nematodes

4·1 Opposite Pig ascarids (nematodes). The males have a curled tail.
4·2 Below Diagram of some monoxenous nematode life cycles. (1) *Trichostrongylus* spp.; (2) *Ancylostoma* and *Necator*; (3) *Ascaris lumbricoides*; (4) *Syngamus*, involving a facultative paratenic host; (5) *Oswaldocruzia* spp.; (6) *Srongyloides* and the heteroxenous cycle; (7) *Stephanurus*.

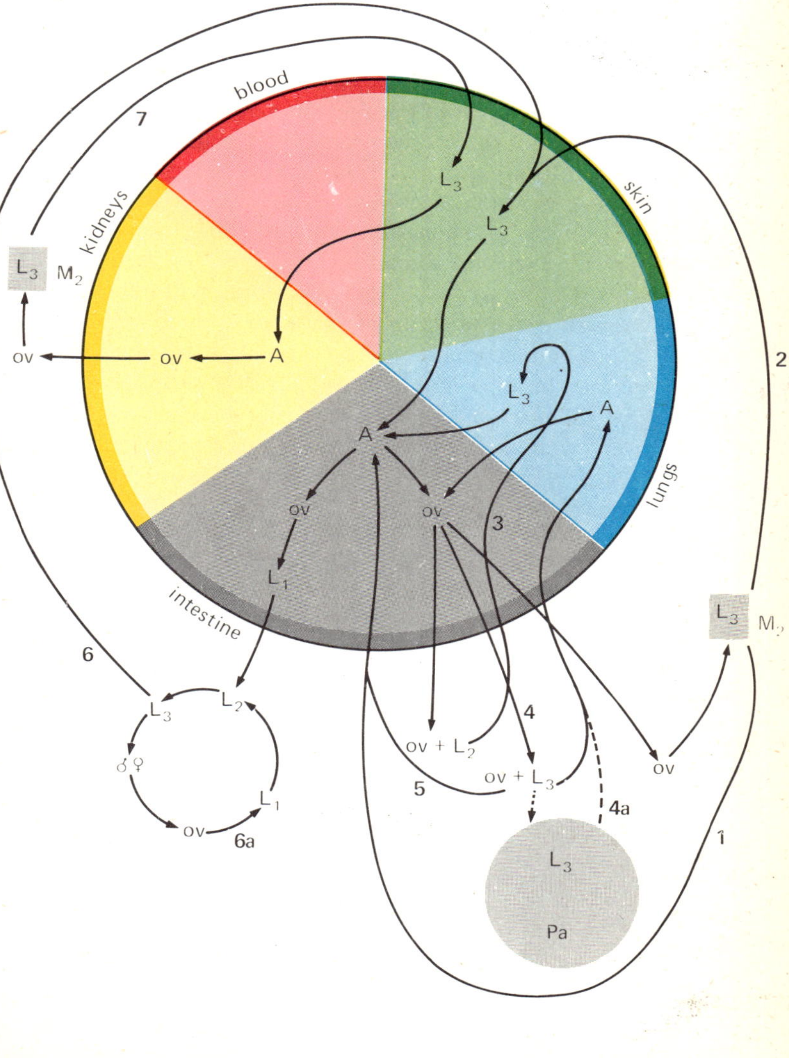

because dust containing third-stage larvae $(L_3)M_2$ is blown over a wide area by the wind.

When the life cycles of the free-living and parasitic nematodes are compared, it is of considerable interest to find that despite complications introduced by the necessity to span several hosts and to undergo migrations within the host, the third-stage larva is still the infestive stage which enters the special ecological niche where the parasite develops into an adult.

When the life cycles of nematodes parasitic in vertebrates are considered from a purely biological standpoint, without taking into account the taxonomic position of the worms, three main kinds of life history can be recognised. In all these cycles the term *definitive host* is used for the host in which the parasite becomes adult and lays its eggs; the *intermediate host* contains the infestive larva and must be eaten by the definitive host or used as a means of introducing the larvae into the final host. The intermediate host is, therefore, indispensible to the completion of the life cycle. There is also a category of host known as the *paratenic host* in which the infestive larvae accumulate without being able to undergo further development; this occurs when the intermediate host is eaten by a host which is not the definitive host and in which the life cycle cannot continue. The paratenic host is not necessary to the fulfilment of the cycle but it can help to disseminate the parasite and under certain ecological conditions could favour completion of the life cycle. A life cycle which involves no intermediate host is termed *monoxenous*, one which does involve such a host is termed *heteroxenous*. There is also a condition where the definitive host can also serve as an intermediate host, as occurs in *Trichinella;* this kind of life history is termed *autoheteroxenous*.

Using these main categories it is possible to obtain an overall view of several life cycles which demonstrate a general tendency towards specialisation.

Monoxenous cycles with a free-living larval stage

(a) With a heterogenous generation:
Under this heading come those primitive life cycles in which a parasitic generation and a generation free-living in the soil are produced successively. For example, in *Rhabdias bufonis*, the parasitic generation which lives in the lungs of frogs and toads is represented by a protandrous, hermaphrodite 'female'. Eggs are voided via the intestine and hatch in humid soil where L_1 larvae can develop in two different ways, depending upon the circumstances. They may give rise either to third-stage larvae L_3 which will invade a new host or to noninfestive third-stage larvae which develop into free-living adult male and female worms. This heterogenous generation lays eggs which hatch in moist earth and produce infestive third-stage larvae. The infestive larvae penetrate via the skin of the host and are carried to the lungs in the blood stream. The same kind of life cycle occurs in *Strongyloides stercoralis*, an intestinal parasite of man.

(b) With no heterogenous generation:
(*i*) *No larval migration in the host:* Third-stage larva protected by the uncast larval cuticle in the soil.

The trichostrongyles of ruminants live in the intestine from where the eggs are evacuated. These hatch on damp ground and the third-stage larvae migrate on to blades of grass where they remain until eaten by the host.

(*ii*) *Larval migration in the host:* Third-stage larva protected by the uncast larval cuticle in the soil.

The hookworms of man and carnivores lay eggs in the gut. These hatch in the soil to produce third-stage larvae. On contact with host skin the larvae penetrate this and are carried to the lungs

Table 3 (a)

MONOXENOUS

Egg	Larva 1 L_1	Larva 2 L_2	Larva 3 L_3	Intermediate host	Paratenic host
Soil	soil	soil	soil	—	—
Soil	soil	soil	soil	—	—
Soil	soil	soil	soil	—	—
Soil	soil	soil	soil	—	—
Soil	soil	soil	soil	—	facultative earthworms insects
Soil	egg	egg	egg	—	—
Soil	egg	egg	lungs	—	—
Soil	egg	egg	intestinal mucosa	—	—
Soil	egg	liver	liver	—	—

in the circulatory system; they spend little time in the lungs and penetrate the alveolar walls, pass up the trachea and are swallowed, completing their development in the gut.

In *Uncinaria lucasi*, a parasite of the Pribilof Island fur seal, this kind of life cycle is remarkably well adapted to suit the ecology of the host. The eggs hatch in the soil and third-stage larvae occur here. On contact with the host these penetrate the skin of the flippers and belly and reach the fat layers and, in the case of female seals, the mammary glands where larvae accumulate in the ducts. Young seals become infested via the milk of the infested females. The worms become mature in fifteen days but at the age of six months

Migration	Adult	Species	Definitive host
—	intestine	*Trichostrongylus* spp.	ruminants
skin-blood-lungs	intestine	*Ancylostoma* spp.	man, carnivores
skin-blood-liver-spleen	kidney	*Stephanurus dentatus*	pigs
intestine-blood-lungs	lungs	*Dictyocaulus filaria*	ruminants
intestine-blood-lungs	lungs, bronchi, trachea	*Syngamus trachea*	birds
—	intestine	*Oswaldocruzia* spp. *Enterobius vermicularis*	amphibians, man
intestine-blood-lungs	intestine	*Ascaris lumbricoides*	man
—	intestine	*Ascaridia galli*	chicken
intestine-liver	liver	*Capillaria hepatica*	rodents

the seals spontaneously eliminate their parasites and do not become reinfested. Thus adult seals are not parasitised by the adult nematodes. The life cycle of the worms can be completed only on land during the seal's breeding season, so this suggests that egg and larval stages must have survived in the soil for at least a year.

In the strongyle parasite of the horse caecum the third-stage larva is swallowed but penetrates the gut wall and enters the mesenteric blood vessels. After about fifteen days the worms mature, migrate back along the capillaries, and regain the gut, where the adult worms become attached.

Stephanurus dentatus is parasitic in the urinary system of pigs, and eggs are laid in the ureters and are eliminated in the urine. Hatching occurs in soil where the resistant third-stage larvae can survive for a long time. After either being swallowed by the pig or having penetrated its skin, the larvae reach the liver and may pass into the circulatory system. They finally leave the liver or spleen and enter the abdominal cavity before penetrating the ureters.

Dictyocaulus filaria lives as an adult in the lungs of many kinds of wild and domestic ruminants. Eggs pass out of the body directly, during coughing or via the intestine, and hatch in the soil. When the third-stage larvae are swallowed by a suitable host they burrow through the gut wall and after a four day migration in the lymphatic system gain the lungs. An unusual method of propagating the third-stage larvae by means of the sporangium of a fungus growing on faecal material occurs in this cycle. The infestive larvae accumulate on the surface of the sporangium and as these are explosive, projecting the spores over the pasture, the larvae are also dispersed. The fungus *Pilobolus* has been shown to throw infestive larvae to a distance of three metres.

Many birds harbour *Syngamus trachea* in their tracheae and lungs. This is often highly pathogenic to young birds reared in hatcheries. The third-stage larva is already formed within the egg when this is laid. The egg either hatches immediately or can remain dormant so that either the egg or the third-stage larvae can be infestive. When swallowed by a bird the larvae migrate out of the intestine and attain the lungs. Occasionally earthworms, molluscs or arthropods swallow the eggs of *Syngamus* and the third-stage larvae accumulate in the body but do not develop any further. These invertebrates thus constitute paratenic hosts which can increase the distribution of the parasites as well as enhance the chances of infestation, since it is more likely that the definitive bird host is more likely to eat an infested paratenic host that to pick up eggs from the ground.

Monoxenous cycles with no free-living larval stage

(*i*) *No larval migration in the host:* In oxyurids of the genus *Oswaldocruzia* living in the intestine of amphibians, the larvae moult twice in the egg so that this contains a third-stage larva which can directly infest a new host swallowing these eggs.

In the case of the human oxyurid *Enterobius vermicularis*, the female leaves the intestine and lays eggs on the peri-anal skin.

(*ii*) *Larval migration in the host:* The eggs of man, horse and pig ascarids are laid in the intestine and reach the ground in a comparatively undeveloped state. They therefore have to remain in the soil until they become mature and contain a second-stage larva. When the egg is swallowed, the second-stage larva is released and bores through the wall of the intestine, to be taken to the lungs in the bloodstream. The third-stage larva develops in the alveoli and from here migrates via the bronchi into the trachea and back down the gut where maturity is reached.

In *Ascaridia galli*, an intestinal parasite of the chicken and other birds, migration of the second-stage larvae which are taken in within the egg is merely a token activity for the larva only buries itself in the mucosa and remains there until the second moult and third-stage larva. This re-enters the intestine where it matures.

The situation in *Capillaria* is remarkably similar to this, considering that this group was derived from a completely different group of free-living nematodes. The same number of moults occur but it is not the third-stage larva that is infestive but the first-stage larva. *Capillaria hepatica* lives in the liver of rodents and its eggs accumulate here without developing further. Death of the host is necessary for the eggs to be released into a dry, external environment where they may develop into a first-stage larva which is retained within the egg. The egg is now infestive and hatches in the gut of a new rodent

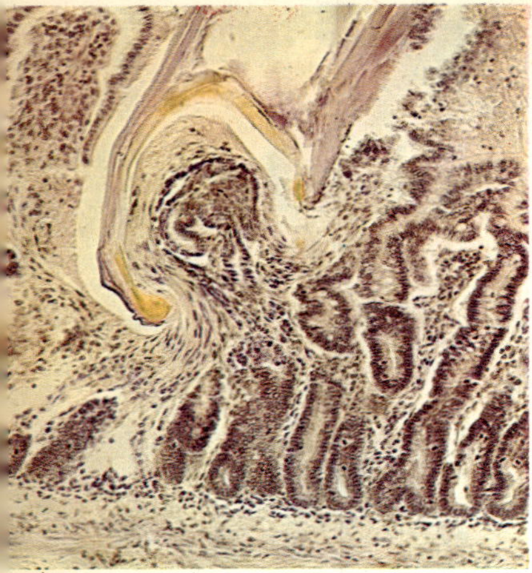

4·3 Left *Ancylostoma* from the dog attached to the intestinal mucosa, showing the function of the buccal capsule.

4·4 Opposite Artificially induced hatching of the eggs of a human pinworm.

4·5 Opposite right *Capillaria splenacea*; section through the spleen of a shrew showing eggs and worms.

host. The larva migrates into the liver. It has been suggested that the cannibalistic tendencies of rodents might aid the infestation of new hosts, but observations do not bear this out, instead showing that the eggs have to spend some time outside the host.

Heteroxenous cycles with a single intermediate host and a free-living larval stage

(*i*) *No larval migration in the definitive host: Habronema muscae* lives in the stomach of horses, the eggs hatch in the intestine and the first-stage larvae pass out in the dung. The larvae are eaten by the maggots of house flies and develop into infestive third-stage larvae which accumulate in the labium of the intermediate host as soon as the adult fly leaves the pupa. The fly is attracted to the lips of horses by their humidity and as soon as it settles there the infestive larvae actively penetrate until they reach the horse's stomach. In tropical regions the fly vector is often attracted to open sores which the nematode larvae penetrate. The worms do not develop here but cause irritation, termed cutaneous habronemiasis, which complicates the injury.

Spiroxys contorta occurs in the intestine of grass snakes and the eggs are laid into water. After a moult in the egg, the second-stage larva hatches and is eaten by a copepod in which it develops into

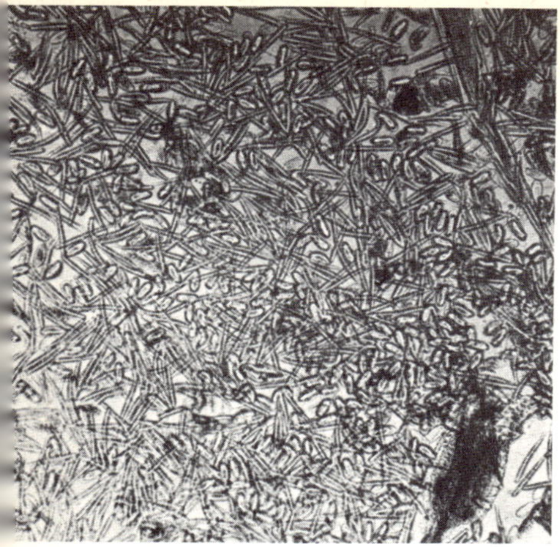 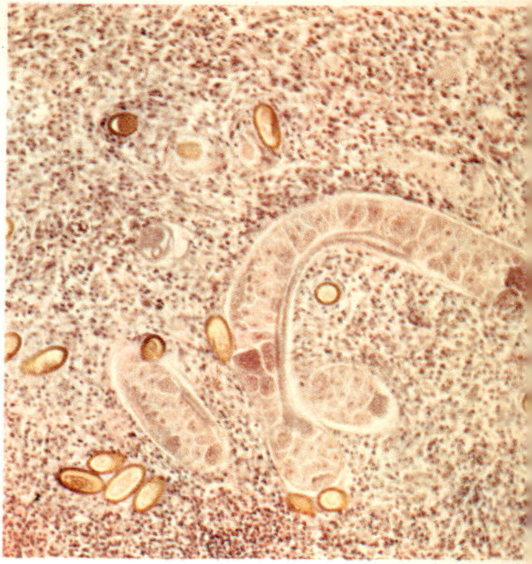

an infestive larva. The grass snake becomes infested by swallowing the copepods. This life cycle can be complicated by the presence of a paratenic host, a fish which eats infested *Cyclops* and is then itself consumed by a grass snake.

(*ii*) *Larval migration in the definitive host:* The protostrongyles are largely lung parasites of ruminants. Eggs hatch in the alveoli and are passed out via the intestine. First-stage larvae penetrate the foot of various molluscs in which the third-stage larva will be formed. The host acquires the infestive larvae by eating infested snails; from the host intestine the larvae migrate to the lungs.

Angiostrongylus cantonensis is a parasite of the pulmonary vessels of the rat and the eggs hatch in the lungs, first-stage larvae passing out of the body by the intestine. They penetrate a mollusc where infestive larvae are formed. Once swallowed by a definitive host the larvae then burrow through the intestinal wall and are swept away in the blood stream to the capillaries of the brain where subadults are formed which migrate to the pulmonary blood vessels. Paratenic hosts may be involved in the life cycle and man can become infested. Cases of tropical meningitis due to the presence of these worms in the brain capillaries have been recorded (Alicata, 1965).

Table 3 (b)

HETEROXENOUS

Egg	Larva 1 L_1	Larva 2 L_2	Larva 3 L_3	Intermediate host	Paratenic host
Stomach	soil	maggot	fly	obligatory	—
Water	egg	water	copepod	obligatory	facultative fish
Lungs	intestine	mollusc	mollusc	obligatory	—
Lungs	lungs, intestine	mollusc	mollusc	obligatory	facultative crustaceans
Uterus	water	copepod	copepod	obligatory	—
Soil	egg	egg	earthworm	obligatory	—
Soil	egg	egg	mouse	facultative	—
Soil	egg	insect	insect	obligatory	—
Soil	egg	insect	insect	obligatory	—
Uterus	blood	insect	insect	obligatory	—
Water	egg	copepod	fish	obligatory	—
Uterus	lymph	blood	striated muscle	also serves as definitive host	—

Migration	Adult	Species	Definitive host
—	stomach	*Habronema muscae*	horse
—	intestine	*Spiroxys contorta*	grass snake
intestine-blood lungs	lungs	*Protostrongylus* spp.	ruminants
intestine-blood-brain	lungs	*Angiostrongylus cantonensis*	rodents, man
intestine-blood-connective tissue	subcutaneous connective tissue	*Dracunculus medinensis*	man, mammals
—	intestine	*Porrocaecum ensicaudatum*	birds
facultative	intestine	*Toxocara canis* *Toxocara mystax*	dog cat
intestine-blood-lungs	air sacs	*Diplotriaena* spp.	birds, reptiles
stomach-oesophagus-lachrymal duct	cavity of eye	*Oxyspirura mansoni*	chicken
blood-connective tissue	connective tissue	*Wuchereria bancrofti Loa loa Acanthocheilonema perstans*	man
—	stomach	*Gnathostoma spinigerum*	
intestine-blood-muscle	intestine	*Trichinella spiralis*	man, mammals

4·6 Diagram of heteroxenous nematode life cycles with a free-living larval stage. (1) *Habronema muscae*; (2) *Protostrongylus* spp.; (3) *Angiostrongylus* and facultative paratenic host; (4) *Spiroxys contorta* and facultative paratenic host; (5) Guinea worm (*Dracunculus medinensis*).

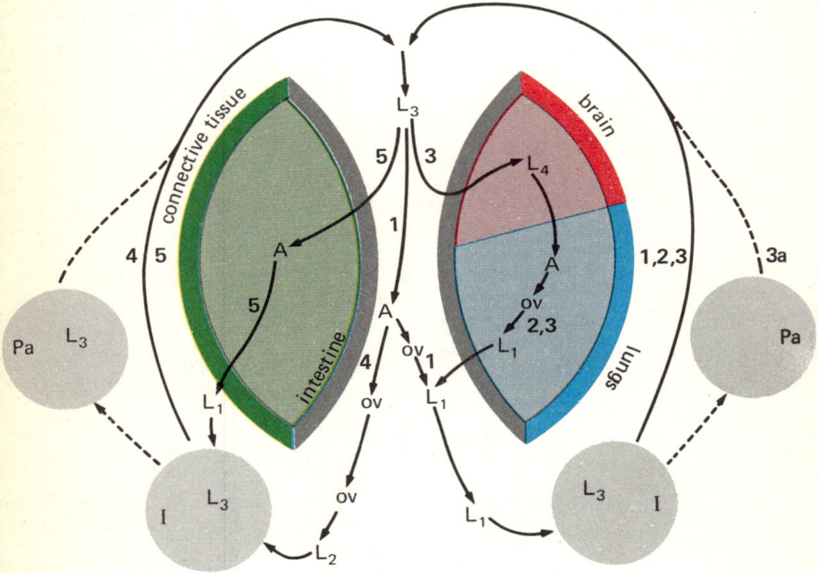

Dracunculus medinensis (the Guinea worm) is a well-known parasite of man and lives under the skin, usually of the leg. The female worm ruptures the covering of skin and emits hundreds of first stage larvae through the break thus formed. These develop into third-stage larvae in the body cavity of a copepod. Man is infested by drinking in copepods with water and the larvae leave the intestine and spend some time in the mesentery. Infestation with the adult worms becomes apparent only after about twelve months.

4·7 Diagram of heteroxenous nematode life cycles which do not have a free-living larval stage (except (6)). (1) *Porrocaecum*; (2) *Toxocara* and transplacental larval migration; (3) *Diplotriaenia*; (4) filaria worm; (5) *Oncocerca*; (6) *Gnathostoma spinigerum*.

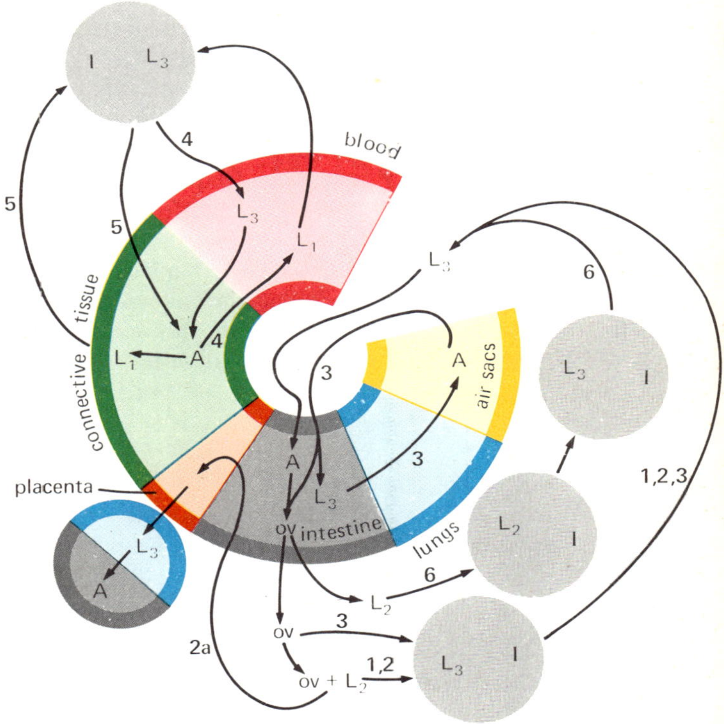

Heteroxenous cycles with a single intermediate host and no free-living larval stage

(*i*) *No larval migration in the definitive host: Porrocaecum ensicaudatum* inhabits the intestine of various birds. Eggs are evacuated on to the ground and the first-stage larva moults within the egg. This, with the enclosed second-stage larva, is taken in to the intestine of an earthworm and here the larva hatches and migrates into the blood vessels where it develops into an infestive third-stage

4·8 The third stage larva of *Porrocaecum ensicaudatum* within the blood vessel situated ventral to the nerve cord of an earthworm.

larva. When the earthworm is eaten by a bird, the larva moults in the intestine and develops into an adult nematode.

The dog ascarid *Toxocara canis* is also an intestinal parasite. The mature eggs contain a second-stage larva. As soon as these are swallowed by a mouse, the larvae hatch in the intestine and migrate throughout the animal, occurring frequently, for example, in the brain capillaries. Dogs acquire the infestation by eating mice. In most cases, however, an interesting alternative to this cycle can occur. Second-stage larvae liberated into the intestine of a dog which has swallowed the eggs traverse the placenta and lodge in the lungs of the foetus. When the puppy is born, the worms enter the intestine and become adult. The first alternative for this life cycle occurs also in the cat ascarid *Toxocara mystax* but transplacental infestation does not occur.

(*ii*) *Larval migration in the definitive host:* There are many species of the genus *Diplotriaena*, which inhabits the air sacs of birds and reptiles. Eggs are voided via the extrapulmonary bronchi and then via the intestine. The infestive third-stage larva develops in an insect which has ingested eggs. This larva, which hatches in the intestine of the definitive host, first reaches the lungs by means of the blood system and then develops to maturity in the air sacs.

Oxyspirura mansoni lives in the lachrymal duct and humour of chickens in tropical regions. Eggs, which leave from the intestine, contain a first-stage larva. When eaten by cockroaches larval development up to the third stage occurs. When infested cockroaches are fed to chickens, the third-stage larva can be recovered from the chicken's eye as little as five minutes after passing through the stomach and up the oesophagus into the lachrymal duct.

The filaria worms of man inhabit the connective tissue of different regions according to the species in question. They are all viviparous and, apart from one exception (*Oncocerca*), the first-

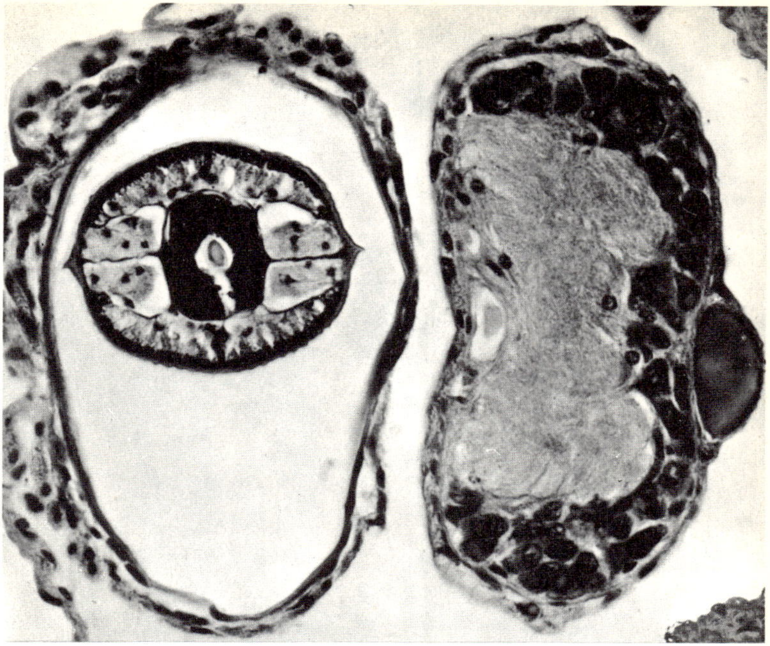

stage larvae, called microfilariae, occur in the blood. Their mass appearance in the blood follows a periodicity appropriate to the particular species of filaria worm. The microfilariae are ingested by haematophagus insects (flies or mosquitoes), and the larvae moult within the bodies of these vectors. The infestive larvae become concentrated in the labium of fly vectors and when these bite a new host the larvae are released on to the skin by rupture of the labium, penetrating rapidly into the body and migrating to the preferred site in which the adult filaria worms occur.

In the case of *Oncocerca*, the adult worms cause nodules of connective tissue to be formed beneath the skin and the microfilariae occur in the dermis of adjacent regions. Adult filariae develop in these regions of the body where infestive microfilariae were inoculated by the vector *Simulium* (the Blackfly). Where microfilariae occur in the eye, blindness may result and this is quite common in certain regions of Africa.

Heteroxenous cycles with two intermediate hosts and a free-living larval stage

(*i*) *No larval migration in the definitive host: Gnathostoma spinigerum* is a parasite of the cat intestine where it often produces tumours. The eggs, laid into water, enclose a second-stage larva which hatches and has to be eaten by a copepod. Curiously enough, however, it does not develop to the third-stage larva here. This stage is reached in a fish which acquires the infestation by eating copepods.

The autoheteroxenous cycle with no free-living larval stage

(*i*) *Larval migration in the host:* This very specialised kind of life cycle, in which the definitive host inevitably becomes the intermediate host, is found only in *Trichinella spiralis*, a parasite of man and a large number of domestic and wild mammals. The adult worms occur in the intestine where the viviparous females burrow into the mucosa and here produce first-stage larvae. These larvae are carried to the heart in the circulatory system and from here are transported throughout the body, but can continue their development only in striated muscle. Here they reach the infestive stage and cause the host tissues to react and enclose the larvae in a cyst. Within this cyst the larvae may remain infestive for several years. A new host becomes infested by eating the flesh of the parasitised host. Transmission of *Trichinella* must originally have been favoured by cannibalism as occurs, normally, among rodents. Man gains the infestation by eating contaminated pork and pigs acquire *Trichinella* by eating pork scraps or occasionally infested rats. Numerous other mammals can act as reservoir hosts in all latitudes, among them being the polar bear in the Arctic and the bush pig in Africa. In temperate regions rats are the main reservoir hosts.

Tables 3a and 3b (pp. 84–5 and 90–1) summarise the main types of life cycles of nematodes parasitic in vertebrates. Although nematodes show little obvious morphological adaptation to parasitism their life cycles are greatly specialised to this way of life. The main kinds of specialisation are the way in which third-stage larvae are produced and that in which the definitive host is invaded, or both. In the most primitive kinds of life cycle, the egg hatches in soil where the two successive larval stages live, but development up to the second- or third-stage larva can also occur within the egg, in which case the eggs do not hatch but have to be passively ingested by the definitive host. In heteroxenous cycles where an intermediate host occurs, this always contains third-stage larvae and becomes infested by eating embryonated eggs or by receiving first-stage larvae (microfilariae). In either case the larvae actively invade the tissues of the intermediate host. In addition to passive ingestion by mouth, transcutaneous infestation can also occur and the third-stage, soil-living larvae penetrate directly into the intermediate (or definitive) host.

The establishment of adult nematodes in the various micro-habitats of the definitive host is often preceded by larval migrations within the body of this host. Even though the routes of these migrations may seem to be determined, this is due to the fact that the larvae are often transported in the circulatory system. The third-stage larvae which penetrate the skin are carried passively to the lungs and the same is true of larvae which bore through the intestinal mucosa, thereby reaching the mesenteric vessels. The genus *Capillaria* is site-specific, however, as is manifest by the localisation of the adult worms. In the white-toothed shrew, five species occur living in the walls of the oesophagus, stomach, bladder and in the liver and spleen respectively. Again migration to the liver and renal system by way of the spleen (*Stephanurus*), movement into the arterioles of the brain (*Angiostrongylus*) or to the air sacs of a bird (*Diplotriaena*), into connective tissue (filariae) or into the eye (*Oxyspirura*) are all active migrations and the nematodes performing them are therefore probably more highly specialised than those attaining their goal passively. The ascarid life cycle shows another kind of specialisation, since in dog ascarids foetal infestation across the placenta occurs in addition to an alternative cycle involving a facultative, intermediate host. Infestation of the litter is thus assured and at the same time infestive larvae can be distributed via the intermediate host to develop into adult worms which will constitute a 'non-litter' infestation. In *Ascaridia galli* the second-stage larva becomes buried in the intestinal mucosa of the definitive host and transforms into an infestive larva which re-enters and attaches within the lumen of the intestine. This life cycle therefore represents an abbreviated migration with the intestinal mucosa assuming the role of intermediate host. Finally, the life history of *Ascaris lumbricoides*, like that of other ascarids of pigs, cattle and horses, involves migration of the second-stage larvae into the lungs where the third-stage larva develops. This life cycle can be considered more specialised than that of other ascarids in the sense that migration within the definitive host is a result of the loss of an intermediate host in which this migration used to occur. It is thus possible to have an idea how the life cycles evolved within a single group of nematodes.

In conclusion it must be remembered that the infestive larvae can be taken in by a paratenic host in which they maintain themselves without undergoing further development. The paratenic host, which is always facultative, assists the dispersal of the nematode into the definitive hosts having the same behaviour. The presence of infestive *Syngamus* larvae in earthworms and insects makes it more likely that the final bird host will become infested than if the only path of infestation was by the bird eating the eggs directly.

The consequences of specialisation in life cycles are examined in chapter 8 where the fundamental problem of host specificity will be dealt with.

5 Flat worms or platyhelminths

The platyhelminth parasites, namely the monogeneans, cestodarians, cestodes and trematodes (digeneans), are too specialised to allow their direct relationship with the free-living forms of turbellarians to be traced. All platyhelminths have in common a body packed with spongy, deformable parenchyma tissue and a closed excretory system consisting of flame cells or protonephridia. They are usually hermaphrodites, separation of the sexes being exceptional. Only these basic features are really comparable because separation of the parasitic forms from the free-living forms happened so long ago that the ancestral lines are no longer visible. In addition the differences between the monogenean-cestode line on the one hand and the trematodes on the other is so great that it seems unlikely that they have arisen from the same stock of platyhelminths. In this account all the groups will be dealt with separately.

Monogeneans

The monogeneans live as ectoparasites on the gills or general body surface of marine and fresh-water fishes. As their size rarely exceeds 6 mm, they would constantly face the risk of being detached were it not for the fact that they are able to fix themselves firmly to their hosts by means of a posterior attachment organ termed the haptor. This is a flattened, often disc-shaped organ which carries a system of hooks. The median hooks may be large and anchorlike and are termed hamuli; peripherally the haptor bears a number (ten to fourteen depending on the group of monogeneans in question) of marginal hooks which arise early in development and are the main larval attachment organs. In certain gill-parasitic monogeneans the haptor has become adapted to holding on to the divided secondary gill lamellae of the host and has itself become divided into a number of suckerlike regions which may become strengthened with hard sclerites to form the so-called clamp attachment organs (figures 5·1a and 5·2b). There is evidence that the progressive development of more efficient attachment organs has been one of the main evolutionary trends in the monogeneans, thereby making them more efficient ectoparasites. Usually the haptor is symmetrical about the median longitudinal axis of the parasite, but it can become secondarily asymmetrical in some gill parasites depending on whether the infesting larva attaches to the gills of the right or left side of the fish (figure 5·3). Thus the body of the parasite comes to conform to the direction of the gill-ventilating current of the host, the attachment organs always being upstream relative to the mouth which is free to browse. Forms with an undivided haptor are termed monopisthocotylineans and are mainly skin parasites; forms with a divided haptor are termed polyopisthocotylineans and are largely gill parasites.

All monogeneans have a mouth, followed by a pharynx which leads into a bifurcate intestine, the two caeca of which may be linked by an intercaecal network; an anus is never present. Because monogeneans move about little on their hosts, and are in general unable to transfer to another fish, their feeding depends to a large extent on their situation, which therefore constitutes a definite microhabitat. Forms living on the skin of the fish and some (monopisthocotylinean) gill parasites feed on mucus; the truly committed gill parasites (polyopisthocotylineans) attach firmly to the gills and feed on blood which they digest after rupturing the branchial capillaries. Associated with these two different kinds of feeding are differences in the nature of the intestinal epithelium of the two groups of parasites; this indicates a long-established specialisation which binds the two forms to their respective microhabitats.

The development of monogeneans is direct and the egg hatches to give an oncomiracidium larva which is equipped with cilia enabling it to swim in search of a new host where it will metamorphose into the adult. These ectoparasitic larvae attach to the host

5·1 Below left **(a)** *Cyclocotyla chrysophri*: haptor with clamps;
(b) *Gyrodactylus*: haptor with hooks.

5·2 Below right Types of haptors with hooks and clamps.
(a) *Ancyrocephalus aculeatus* in ventral and lateral views;
(b) sagittal section through a clamp showing its mode of action.

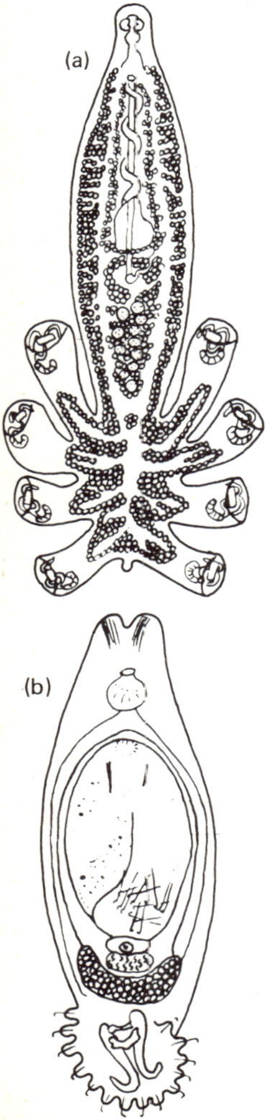

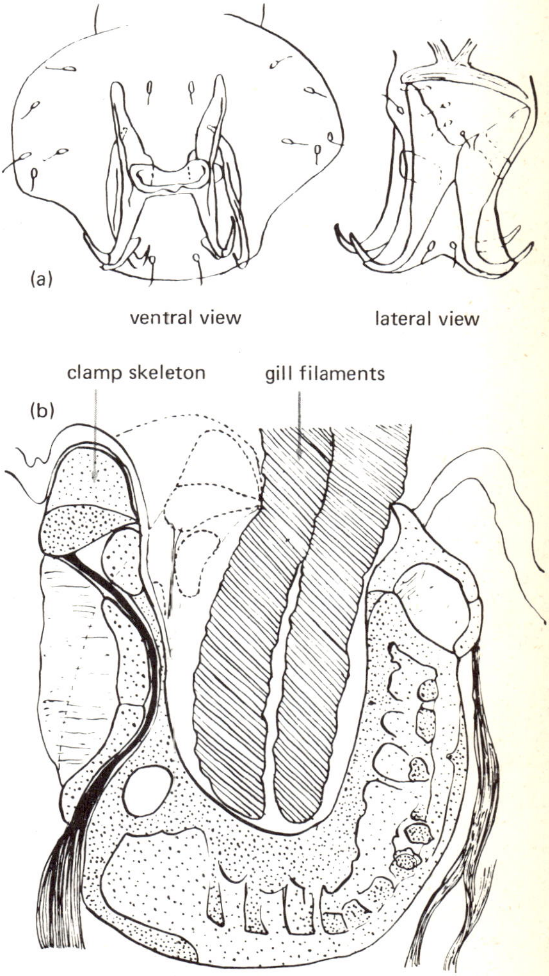

by means of the haptor immediately they contact the skin or gills. The larval haptor is often quite different from that of the adult, especially in the case of the polyopisthocotylinean gill parasites.

The eggs can be remarkably large relative to the adult and they are frequently equipped with one or more filaments which are extensions of the egg envelope. These filaments are not, in fact, used to fasten the egg to the host but tangle with the filaments of other eggs which are laid, often very rapidly, sometimes in large batches. Held together by these appendages the eggs form small bunches which, owing to their low density, either float at the junction of two water layers or sink to the bottom. It is possible that in some cases currents created by swimming fishes could re-circulate these egg masses so that they were taken in with the respiratory current of the fish. If the fish was a suitable one, the eggs fastened to the gills might then hatch and the larvae would either attach to the gills or, in the case of skin parasites, could be transported further in the opercular current to the skin of the host, where they would then attach.

Even though many larval forms have been described recently, we are still far from understanding the life cycles of many mono-geneans, especially those parasitising marine fishes. Experimental work with marine fishes does pose several problems however, for apart from difficulties concerned with maintenance or inhibition of normal behaviour patterns such as migration or shoaling, the fish may be infested from the outset, thus complicating the results.

Most cyprinid species, especially the carp, tench, roach and crucian carp which occur in fresh water, harbour on their gills monogeneans belonging to the genus *Dactylogyrus*, which are not generally harmful. However, many species of cyprinid are reared commercially in ponds – that is, in a relatively confined space where the possibility of a massive infestation is high – and this can affect the host adversely.

5·3 Example of an asymmetrical haptor (*Axine*).

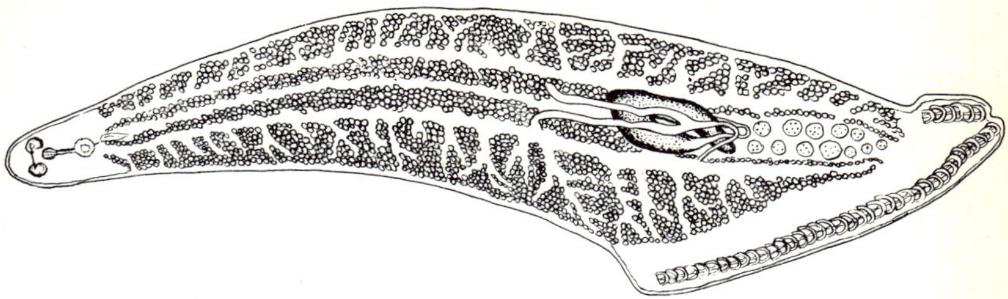

5·4 Below Life cycle of *Dactylogyrus*. (**a**) Larva within the egg; (**b**) larva hatching; (**c**) young worm attached to a fish; (**d**) adult worm.

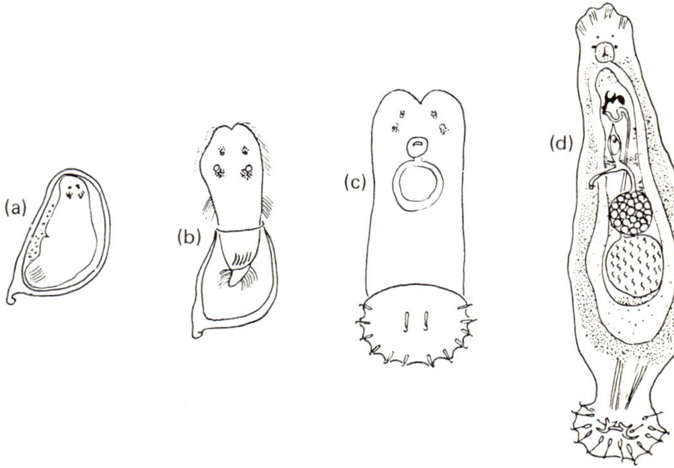

The egg of *Dactylogyrus vastator* hatches at the end of four days when the water temperature is 20°C. The rate of larval development is proportional to the ambient temperature when this is in the range of 20°C. The larva possesses three bands of locomotory cilia, also four photosensitive pigment spots; it is capable of surviving for six to eight hours, but if after this time a host has not been located the larva dies. Arriving on the gills of the fish the larva loses its ciliary plates, attaches and grows to the adult form

5·5 Life cycle of *Diplozoon paradoxum.* (**a**) Egg with filament partially unwound; (**b**) oncomiracidium; (**c**) diporpa larva; (**d**) two diporpa larvae joined by their sucker; (**e**) fused larvae; (**f**) adult.

(figure 5·4c). The whole duration of the cycle is not more than ten to twelve days, after which period the adult dies. During this time, however, sixteen new individuals will have been produced which in their turn will give rise to 256 successors before dying. The reproductive rate of these parasites is obviously fairly high, so this gives an idea of their potential influence on fishes reared in confined spaces where, from an economic standpoint, often catastrophic losses occur.

Another genus, *Gyrodactylus spp.*, also lives on the gills of a whole series of fish that are farmed (notably trout), and is characterised by the absence of a larva, the genus being viviparous. In this case the viviparity is complicated by an unusual kind of polyembryony. In fact the zygote divides into four groups of cells, each of which gives rise to a larva but with the peculiarity that the larvae are enclosed one within the other (like a Chinese puzzle box). Before dying each mother worm gives birth three successive times to these 'quads' and each of these reproduces at the same rate. It has been calculated that in thirty days *Gyrodactylus* will have produced 2,452 individuals. As all the individuals are produced *in situ* on the gills of the host, there is a danger that asphyxia will rapidly develop. Heavily infested fishes are rarely encountered in nature, however, so it must be concluded that the crowded conditions in fish farming are responsible for such massive infestations, acting either to reduce the physical resistance of the fish or by the crowding effect favouring enhanced contact between fish and promoting the possibility of exchange of their external parasites.

On the gills of certain cyprinid fishes there occurs a very interesting monogenean showing a biological feature unique in the animal kingdom in that the 'body' is made up of two fused individuals each of which is unable to survive alone. The eggs of this worm have a long fine polar filament by which they become entwined, forming small bunches which float in the water and then eventually

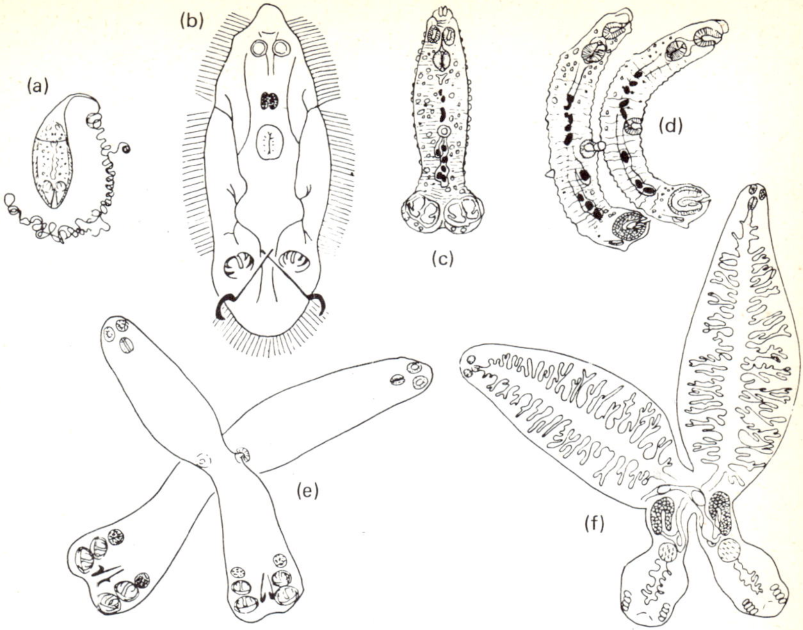

sink to the bottom. The larvae have three plates of cilia arranged at each side of the body and there is a single eye spot. The larval haptor is armed with two long jointed hooks and a pair of clamps is already formed (figure 5·5b). In the absence of the host the larva will survive for about six hours before dying. If the larvae are taken in with the inhalent current, they will attach to the gills and lose their cilia. Two new pairs of clamps appear and a small circular sucker develops in the midregion of the ventral surface of the larva, while on the dorsal side, at almost the same point, appears a small button formed by a bulge in the outer layers. The larva has a straight-sided, sac-like intestine and can live on the gills of its host for several weeks by feeding on blood, but without developing further. This larva is called a *diporpa;* they were described long ago by authors who did not realise that they formed a stage in the life cycle of *Diplozoon*. When two diporpae come into contact each grasps the dorsal button of the other by means of its ventral sucker (figure

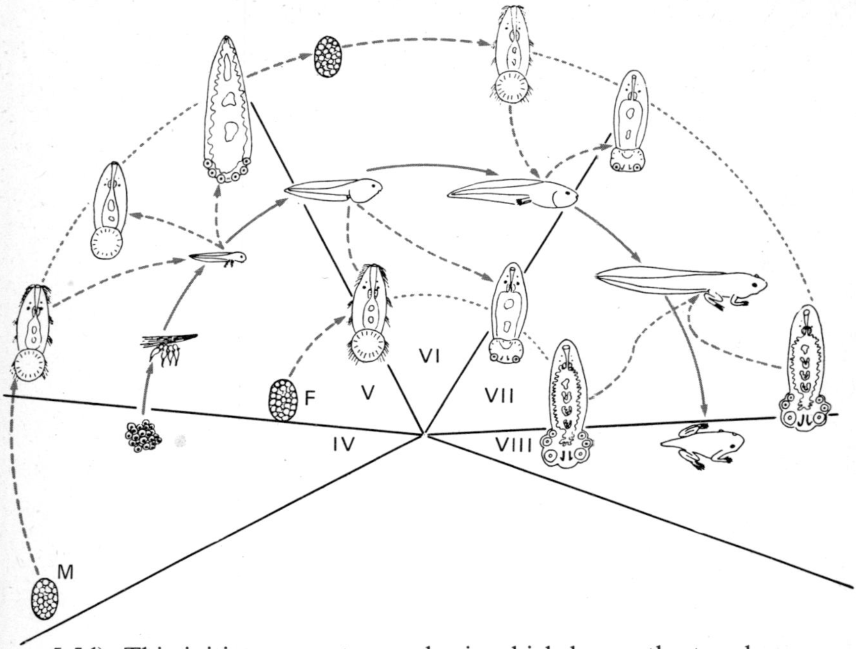

5·5d). This initiates a metamorphosis which leaves the two larvae fused at the level of their mutual attachment points. The intestine ramifies and anastomoses with that of the partner so that food absorbed by one is shared by the other. The male and female genital ducts also become reciprocally fused so that the two individuals are in permanent copulation and cross fertilisation occurs despite the fact that each of the partners is a hermaphrodite and potentially capable of self-fertilisation.

The developmental cycle of the frog polystome is particularly interesting because it illustrates how a form living on the gills of an aquatic vertebrate has become an endoparasite of a terrestrial vertebrate. *Polystoma integerrimum* inhabits the bladder of the frog (*Rana temporaria*). Under the influence of hormones, which appear in the urine of the frog at the breeding season in spring, *Polystoma* lays its eggs which pass out when the frog enters the water to breed. As a general rule male frogs enter the water several

5·6 Life cycle of *Polystoma integerrimum*. (M) Eggs laid by the parasites of male frogs which enter the water in early April; (F) eggs laid by the parasites of female frogs which join the males in late April. The larvae hatching from the first batch of eggs find only very young tadpoles, on which they become neotenic. Larvae from the later eggs coincide with older tadpoles and develop directly into adults when the tadpoles metamorphose in August.

days before the females so that polystome eggs produced by the males have already started to develop and are more advanced than those laid by parasitised female frogs. When the first tadpoles hatch, the polystome larvae attach to their gills, which are still external, and feed on the blood of the host. When the external gills disappear and the branchial region of the tadpole becomes covered with skin to form the branchial chamber, the polystome larvae are also enclosed. These larvae on young tadpoles undergo accelerated development of the reproductive organs relative to the definitive haptor and other organs. Neotenic individuals are therefore produced which retain larval characters but have acquired the ability to reproduce. The uterus of the neotenic forms encloses only a single egg which is then laid and the resultant ciliated larva penetrates the spiracle of older tadpoles but is not usually neotenic. When the tadpole begins to metamorphose, the hind limbs appear first. The anterior limb buds then begin to form. These are covered by the skin of the branchial cavity but this eventually ruptures and the forelimb is protruded through the hole so formed. It is through this opening that the polystome larvae leave to loop along the body of the young frog in order to penetrate into the bladder via the cloacal opening. Young frogs are therefore infested with polystomes during the tadpole stage, on which the larvae live as ectoparasites before passing into the bladder to become endoparasites.

It might appear that, because the life cycles of monogeneans are purely aquatic and complete themselves in a relatively limited environment, infestation of new hosts by the larvae will be relatively straightforward. This, however, may not be the case in forms parasitising marine fishes. Unfortunately these are still incompletely understood and research carried out on fishes in aquaria is not easily extrapolated to fit conditions in open water. In a number of bottom-living forms such as rays and torpedoes, which often harbour monogeneans, the way in which the cycle is completed may resemble conditions found in fresh water. *Acanthocotyle lobianchoi* lives on the ventral surface of the ray *Raia montagui*; the eggs fall to the bottom and, due to a special secretion, stick to sand grains. The larva lacks cilia and is therefore not very motile. Rays are infected while lying on the sea bed, where they come into contact with the larvae (Kearn, 1967). However, in the case of teleosts and sharks of open waters, the completion of the cycle, especially incubation of the egg and host-finding by the larva, must pose great difficulties. The sea is not a homogenous body in terms of its physical and chemical features but is often composed of layers which differ sufficiently from surrounding regions to form distinct habitats with characteristically different fauna. The fish occupy those zones where they find their food and which have, among other things, a suitable temperature and salinity. The eggs of marine monogeneans often have filaments and tend to be laid in bunches. They become entangled by their filaments and might perhaps float between water layers where the density of the water prevents them from sinking. Thus distinct ecological territories could be formed where fish would encounter monogenean larvae or where established monogeneans would lay their eggs. It is not surprising that fishes living in shoals – herring, tunny and cod – are more often parasitised than others.

Cestodarians

The cestodarians are platyhelminths which bear a certain superficial resemblance to the Monogenea but which live in the intestine or in the body cavity of their hosts. Although lacking a gut they are different enough from the cestodes to be dealt with separately, contrary to former practice. The cestodarians consist of two groups, the gyrocotylideans and amphilinids, which differ not only morphologically but also in their general biology. The former live in the intestine of a group of archaic fishes, the holocephalans,

5·7 Hypothetical life cycle of *Gyrocotyle*. (**a**) Adult worm; (**b**) egg containing larva; (**c**) freshly hatched ciliated larva; (**d**) larva inside blood vessel; (**e**) young larva free in the gut having lost its ciliae.

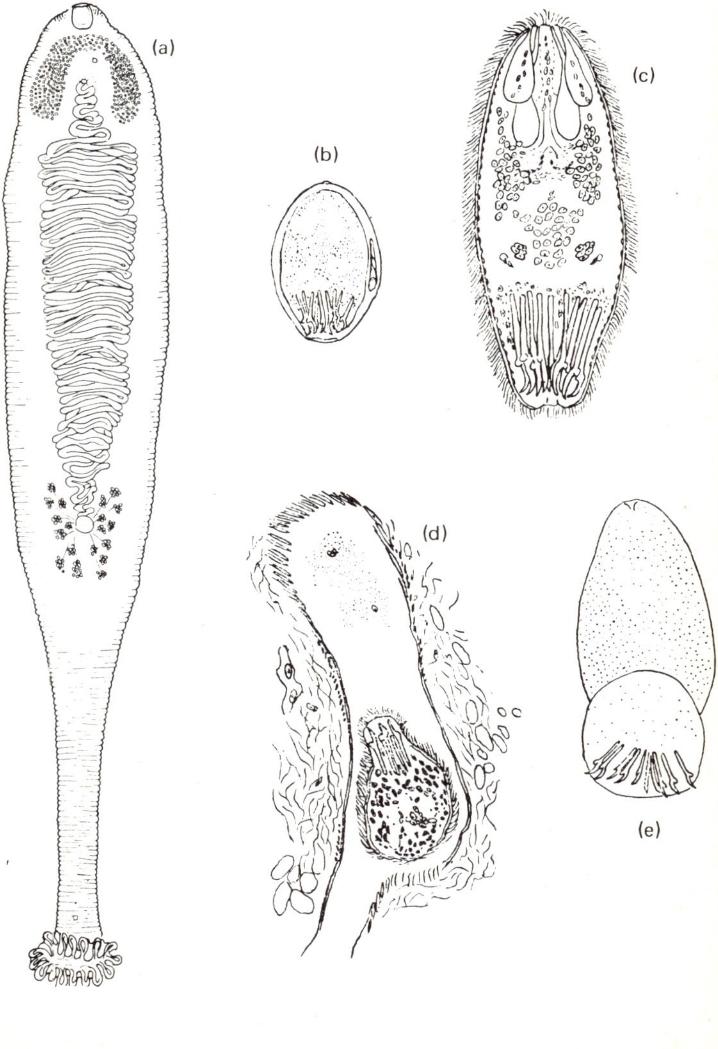

5·8 *Amphilina foliacea*.
(**a**) General view ; (**b**) anatomy ;
(**c**) lycophore larva newly emerged
from the egg showing the
conspicuous head glands.

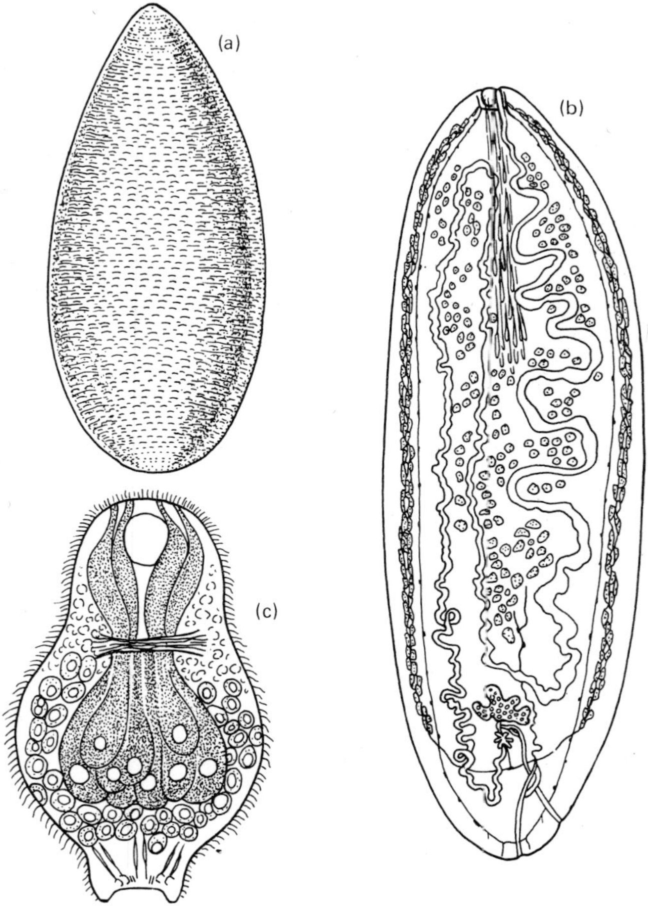

represented by the chimaerids, and the latter live in the body cavity of fishes and occasionally in aquatic tortoises.

The gyrocotylids are about 1–3 cm long and have a kind of funnel at the posterior end of the body, the edges of which are often highly folded and which serves as the attachment organ, known as the rosette organ, in the spiral valve of the host. At the other end of the body is a depression which might represent the vestiges of a larval penetration organ. The genital openings, for instance the uterus, are also situated in this region. The eggs are small and numerous, with a thin shell. They enclose a ciliated larva, the *lycophore*, which is formed before the eggs are laid. This hatches spontaneously in sea water. Anteriorly it has a large gland mass formed of discrete gland elements which all open on the tip of the anterior end. Posteriorly, the larva bears ten pairs of hooks. These larvae are often found in the host intestine apparently having hatched prematurely and survive for several hours in sea water. The life cycle is not known but experimental work by Manter (1951) provides clues as to how this might occur. He observed that the larvae would penetrate into isolated pieces of intestinal mucosa and would be found mainly in the blood capillaries. It is therefore possible that the normal route of infestation is via the gill surface, the larvae attaching by means of their hooks before penetrating the capillaries, using the anterior glands. In holocephalans, blood from the dorsal aorta passes into the intra-intestinal artery, which supplies the free border of the spiral valve, so lycophores might use this route to the intestine which would finally be gained by perforation of the artery wall. It may be significant that in the very young larvae discovered by Manter in the gut (figure 5·7e), the anterior glands have been exhausted, perhaps after passage from the intra-intestinal artery into the lumen of the intestine. The ten posterior hooks of this larva are disposed posteriorly on a kind of lappet (the cercomere) which is the primordium of the future adhesive organ (the rosette). This therefore bears a certain resemblance to the haptor of monogeneans.

The amphilinids are flattened and often transparent, they measure from 6 to 28 cm in length and occur in the body cavities of fresh-water fishes and in an Australian fresh-water tortoise. Only *Gigantolina* lives in a marine fish. The anterior part of the worm is occupied by a large protractile muscular rostrum associated with a well developed system of glands. As will be seen, this acts as an organ of penetration. The uterus opens at the base of this rostrum (figure 5·8). Parasites occupying the body cavity of their hosts have to face the difficulty of getting the eggs to the outside. In this case worms living in the body cavity of the sturgeon have no problem because the eggs leave via the abdominal pores. When a silurid fish is infected with *Amphilina*, however, perforations in the body wall at the level of the pectoral fins are present and as the anterior end of the amphilinid is often found in close association with one slit, it seems likely that this is created by the action of the glandular rostrum. When amphilinids occur in the coelomic cavity of a tortoise, worms and eggs can be found in the lungs, which suggests that the normal route by which eggs leave the host is through the respiratory tract of the host. The eggs already contain a ciliated lycophore larva with prominent glands when they are laid. The eggs have not been seen to hatch in water, which makes it likely that an intermediate host is involved in the life cycle. The only life history so far known – and that only imperfectly – is that of *Amphilina foliacea*, a parasite of the sturgeon. The eggs are eaten by fresh-water gammarids and amphipods and the lycophore burrows into the body cavity of the intermediate host, developing after thirty to sixty days, depending on the temperature of the water, into a juvenile about 4 mm long which already possesses genital primordia. It seems likely, although this part of the cycle has not been confirmed experimentally, that parasitised gammarids are eaten by a sturgeon and that the larvae are released in the gut and bore through the mucosa into the body cavity where they become adult.

There are therefore probably two kinds of life cycles in cestodarians, one with direct infestation of a definitive marine host, as may occur in *Gyrocotyle*, and another with an intermediate host for fresh-water forms, such as *Amphilina*. In both cases the hosts are 'primitive' kinds of vertebrates.

Cestodes

The cestodes are particularly highly adapted intestinal parasites occurring in all groups of vertebrates except crocodiles. Flat and 'tape-like', they have a continuous zone of proliferation which gives rise to the proglottids, in front of which is situated an often complicated adhesive organ, the *scolex*. The 'segmented' body is termed the *strobila*, and contains reproductive organs at all stages of development, each proglottid having separate male and female organs. The further the proglottids become from the growth zone, the more advanced their stage of sexual maturity so that the hindmost proglottids contain only masses of eggs. These ripe proglottids are continuously detached and voided from the intestine, to be replaced by growth at the anterior end, so that the length of the worm, estimated from the total number of segments, hardly varies. These worms have no gut but electron microscope studies on the outer tegument show that in fact these parasites have their intestine investing the outside of their bodies! They are covered by a syncytial epidermis bearing regular villus-like microtriches, which are strengthened at the tip and may be important in increasing the surface area available for the absorption of nutrients. The epidermis itself is a syncytial cytoplasmic layer, the cell bodies of which are situated in the parenchyma beneath the basement lamina. So

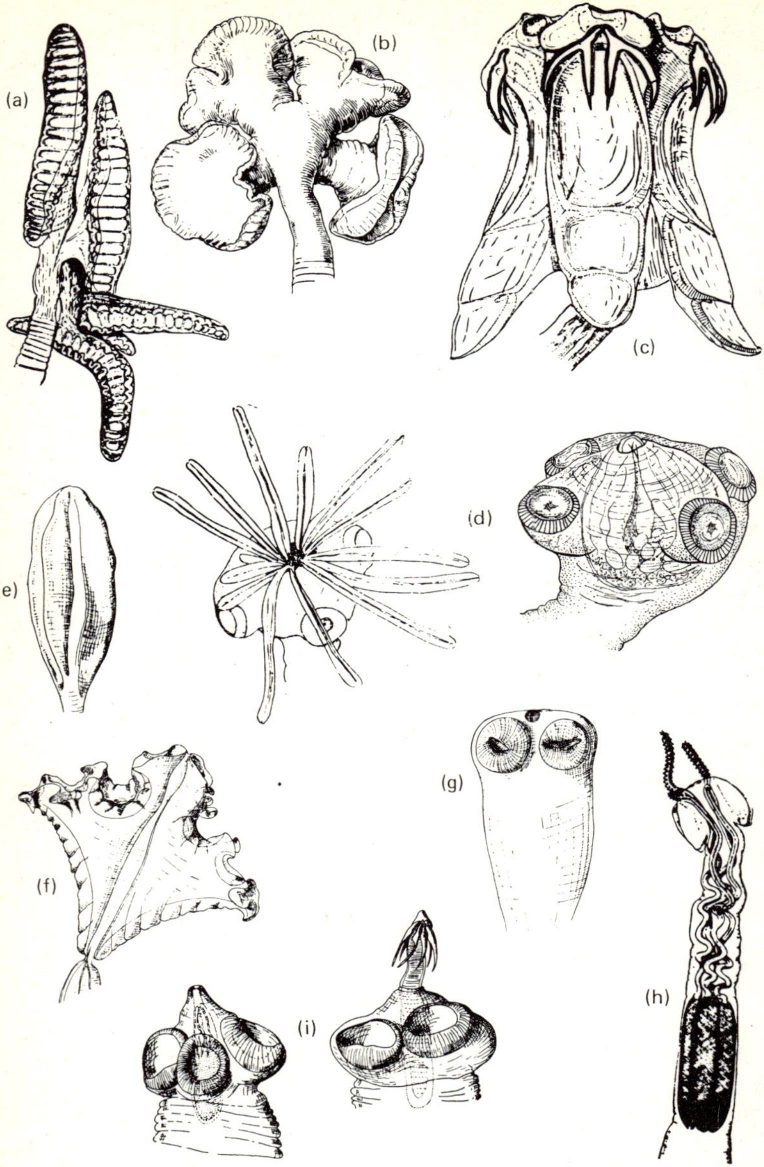
(a)
(b)
(c)
(d)
(e)
(f)
(g)
(h)
(i)

5·9 Opposite Types of scolices in cestodes. (**a**) *Rhinebothrium*; (**b**) *Phyllobothrium*; (**c**) *Acanthobothrium*; (**d**) *Polypocephalus* tentacles retracted and tentacles completely evaginated; (**e**) *Diphyllobothrium*; (**f**) *Duthiersia*; (**g**) *Ichthyotaenia*; (**h**) *Tetrarhynchobothrium*; (**i**) *Hymenolepis*.

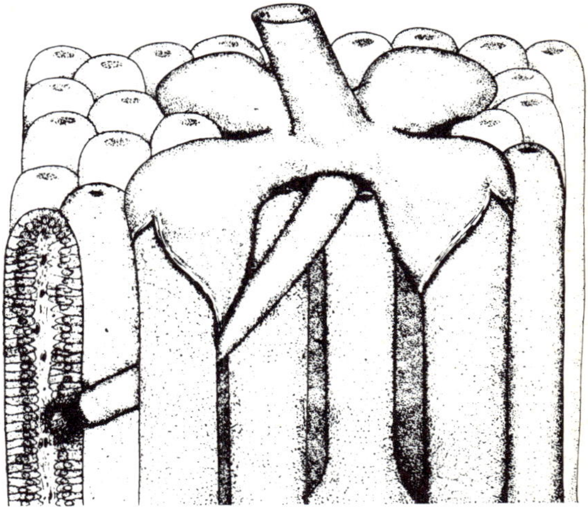

5·10 Above Scolex of *Pseudanthobothrium* attached to the intestine of a ray. Each bothridium is individually attached to a villus and the myzorhynchus attaches to a separate villus.

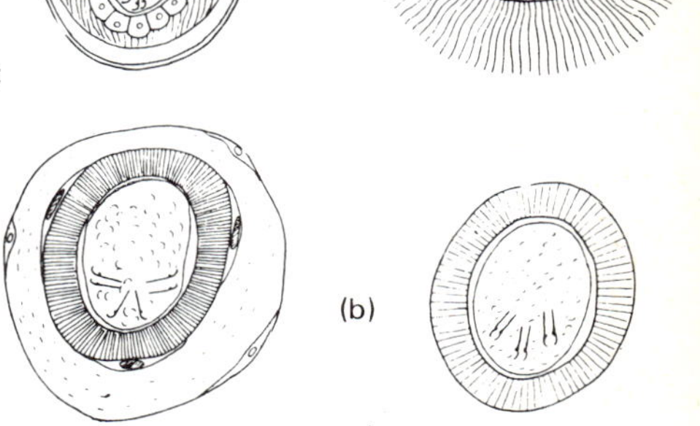

5·11 Right (**a**) Egg and ciliated oncosphere of a pseudophyllidean; (**b**) egg and oncosphere of a cyclophyllidean surrounded by the embryophore.

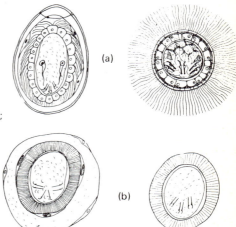

5·12 Types of infestive larvae in cestodes. (**a**) Oncosphere; (**b**) procercoid; (**c**) plerocercoid; (**d**) cysticercoid; (**e**) staphylocyst; (**f**) hydatid; (**g**) coenurus; (**h**) cysticerus.

this living outer layer might well be compared with cells of an intestine as it is able to absorb nutrients and to synthesise substances indispensable to the rapid metabolism associated with continuous growth of the strobila.

The scolex is a symmetrical structure, armed with two to four attachment organs which may be highly complicated and may be adapted to the particular nature of the intestinal mucosa. The most intricate adhesive organs occur in elasmobranch tapeworms which have four suckers with muscular edges, which may be sculptured in such a way as to conform to the pattern of the villi on the wall of the spiral valve (figure 5·10). Hooks are present in some species to facilitate attachment (figure 5·9c) and sometimes also tentacles which penetrate the mucosa (figure 5·9d). The tetrarhynchs are armed with four protractile tentacles armed with spines which anchor the worm in the thick mucus layer covering the surface of the spiral valve. In the tapeworms parasitising terrestrial vertebrates, the structure of the scolex is simplified and in general is reduced to four suckers (figure 5·9g). There may also be a muscular retractile rostellum which bears one or two circlets of hooks aiding the attachment of the scolex to the folded intestinal mucosa (figure 5·9i). Each proglottid encloses male and female reproductive organs which develop independently of those of other proglottids but the development of which is related to their position along the strobila, as has been mentioned. The eggs can either be liberated independently in the intestine or may be eliminated within the gravid uterus when the proglottids detach from the strobila. Mature eggs enclose an oncosphere, or hexacanth larva, so called because it is armed with three pairs of hooks which are used to penetrate into the tissues of the intermediate host (figure 5·11). We are far from understanding the life cycles of all cestode groups, especially in the case of those parasitic on marine fishes, which are not susceptible to experimental investigation. Biologically speaking,

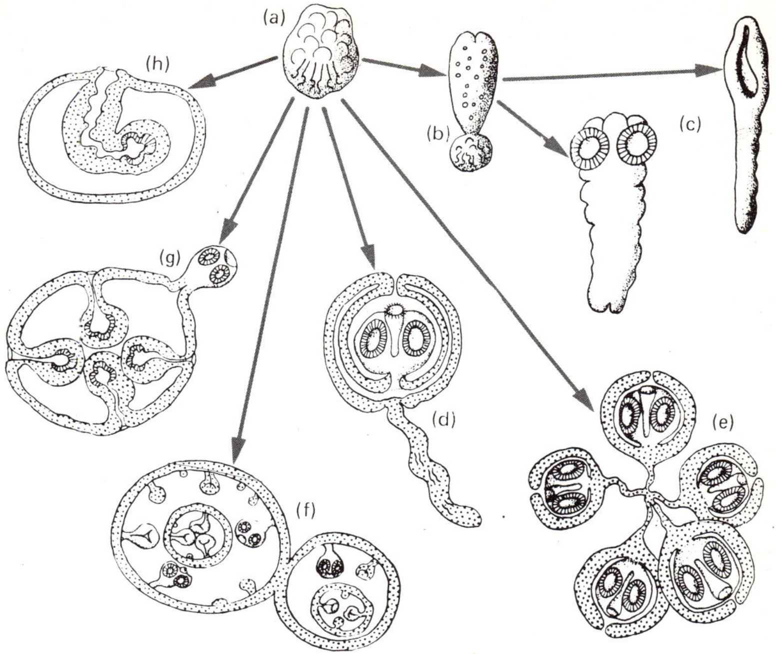

two main kinds of life cycle can be distinguished, according to whether the intermediate hosts are aquatic or terrestrial. Paratenic hosts are especially a feature of aquatic life cycles.

The structure and nomenclature of cestode larvae is fairly complicated but a number of definite stages can be recognised and used in guidance. The procercoid larva occurs in all cestode life cycles but only appears as a distinct phase in aquatic cycles. In terrestrial life cycles this only represents a stage through which the larva passes and it is difficult to demonstrate unless abundant material is available. All cestode larvae, when fully formed and infestive, already possess an adult scolex, by which they can often be identified. Although there are exceptions to this, due to retarded development, this does hold true in most cases. The different kinds of larval envelopes are physiologically homologous in the sense that they serve the common trophic function of nourishing the

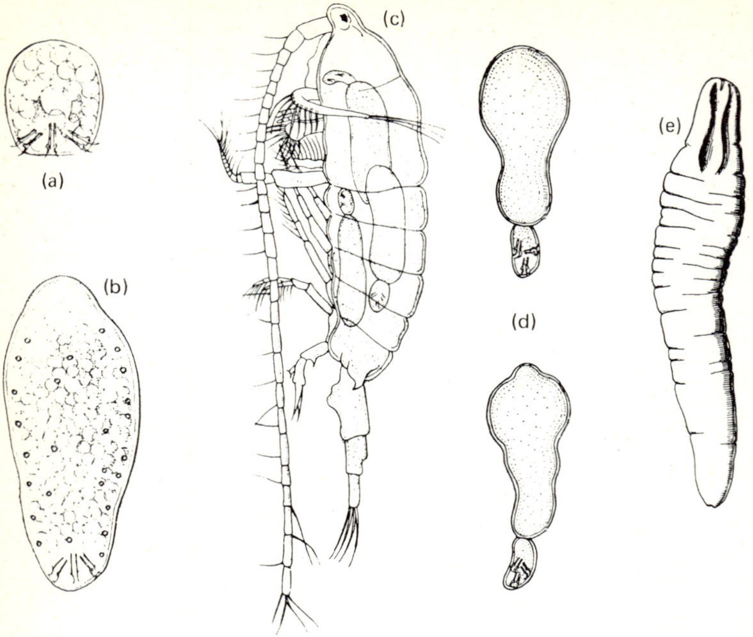

larva in the intermediate host. The structure of these envelopes can be used in the classification and nomenclature of the larvae; they disappear as soon as the larva reaches the intestine of the definitive host where the scolex becomes attached to the intestinal mucosa.

Three basic types of infestive larval stage exist: the plerocercoid, the cysticercoid and the cysticercus (figure 5·12). The first occurs in either vertebrate or invertebrate intermediate hosts in aquatic life cycles, the second in invertebrate intermediate hosts in terrestrial cycles and the third, the cysticercus, in vertebrate hosts involved in terrestrial cycles. The plerocercoid itself lacks a larval envelope but the posterior part is often enormous, as in the tetrarhynchs, and can absorb nutrients. The scolex of the cysticercoid is not invaginated but merely drawn into a pit at the anterior end and the walls of this constitute the larval envelope, while the scolex of the cysticercus is

5·13 Life cycle of *Diphyllobothrium latum.*
(**a**) Oncosphere; (**b**) early stage in the formation of the procercoid;
(**c**) copepod harbouring three procercoids in the haemocoel;
(**d**) living procercoids; (**e**) plerocercoid larva taken from
the second intermediate host (a fish).

invaginated into a closed vesicle containing fluid resulting from interactions with the host. Multiplication of the scolex by asexual larval reproduction can occur in all three types but is uncommon in plerocercoids. It reaches its peak in *Echinococcus* where larval multiplication in a favourable host could theoretically continue indefinitely.

(a) Aquatic cycles:
(*i*) *With two intermediate hosts and a facultative paratenic host:* The classic example of this kind of cycle is *Diphyllobothrium latum*, a parasite of fish-eating mammals and man in the northern hemisphere. The egg hatches in fresh water and liberates an oncosphere which has a ciliated coat which keeps it moving. This larva, termed a *coracidium*, is eaten by a copepod and once in the intestine sheds the ciliated coat, using the three pairs of hooks to penetrate into the body cavity of the crustacean where it develops into a procercoid about 300 μ long with a body constricted into two regions (figure 5·13): an anterior well-differentiated region which will give rise to the future larva and a posterior portion which still bears the six oncosphere hooks. The development of the procercoid then ceases and it can survive within the copepod as long as this remains alive. When a copepod harbours many procercoids it tends to sink to the bottom. Development of the procercoid recommences when the copepod is eaten by suitable fish, usually a perch, although other species can also harbour infestive stages. The procercoids are liberated in the stomach or intestine of a fish and, having burrowed through the gut wall, become lodged in the muscles, especially those of the lateral body region. After several days the procercoid develops into an infestive larva known as the *plerocercoid* which already has a formed scolex, and an elongated body, 10–15 mm long and is of a milky white colour due to the presence of numerous calcareous corpuscles in the parenchyma. The posterior end bearing

the six larval hooks has by this time disappeared. This plerocercoid stage is infestive for as long as the fish remains alive and develops into an adult three weeks after the infested fish is eaten by a suitable definitive host. The total span of the life cycle is approximately three months, provided no paratenic host intervenes. An example of this would be when a carnivorous fish such as a pike devours a parasitised perch. The larvae escape digestion and move out of the pike's intestine where they remain in the body cavity and can survive there without either growing or differentiating further. A pike aged twelve years or so can accumulate hundreds, even thousands of larvae, which are thus removed from the cycle leading to infestation of man since only wild mammals eat the viscera of pike.

Diphyllobothrium mansoni is a parasite of wild carnivores and of the dog, especially in the far east but also parts of southern Europe. The life cycle is similar to that mentioned except that the second intermediate host is usually an amphibian or a reptile (snake or lizard) which harbours the plerocercoids in its muscles. The involuntary intermediate paratenic host in this case is man, and this is due to a local custom whereby a freshly killed frog is used as a poultice to remedy conjunctivitis. The plerocercoid larvae migrate out into the conjunctival fold and cause 'ocular sparganosis'.

Large African carnivores often harbour *Diphyllobothrium theileri*, which has an unusual life cycle in that the second intermediate host is a mammal which becomes infested by drinking water containing copepods harbouring procercoids. Man acts as one of these intermediate hosts and might well be thought to form a cul-de-sac for the life cycle. Certain funeral customs, however, allow the life cycle to be accomplished. The nomadic Masai tribe show a high degree of subcutaneous infestation with plerocercoids. They bury their dead only shallowly or even place the corpse on

5·14 Life cycle of *Ligula avium*.

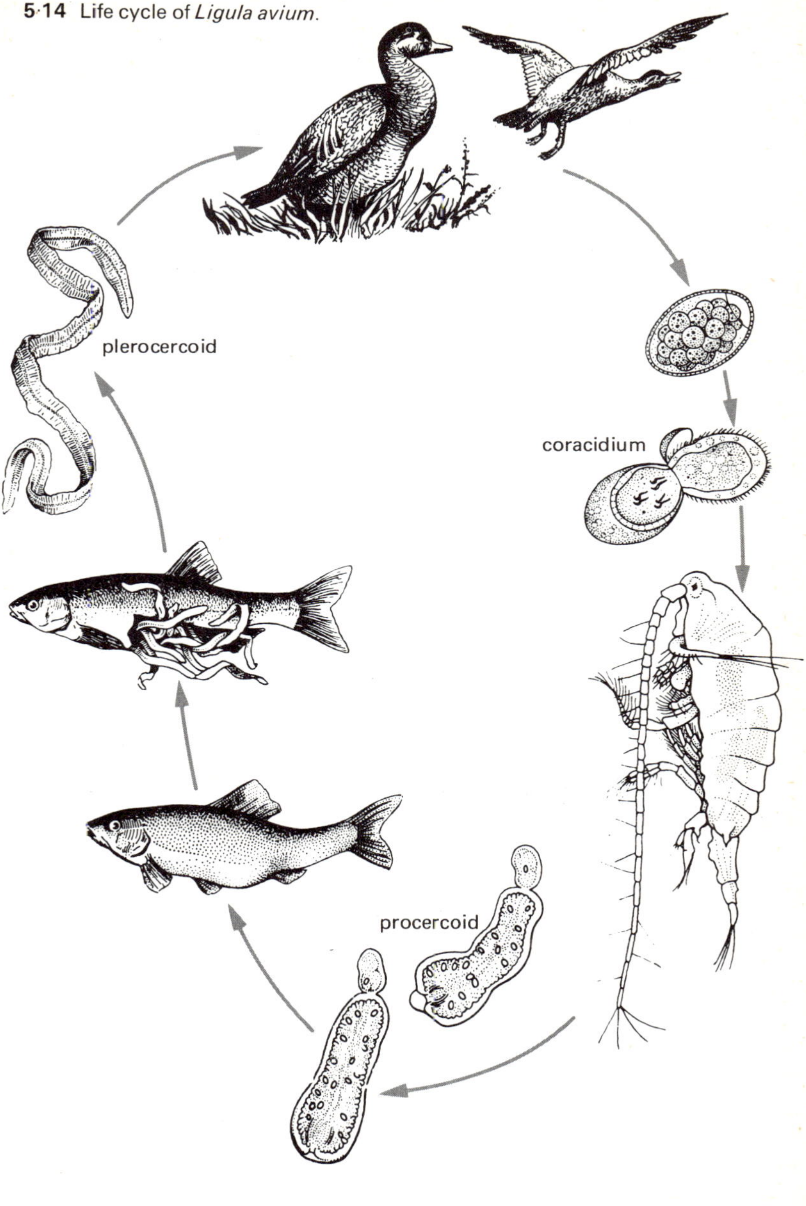

5·15 Life cycle of *Ophiotaenia racemosa*.

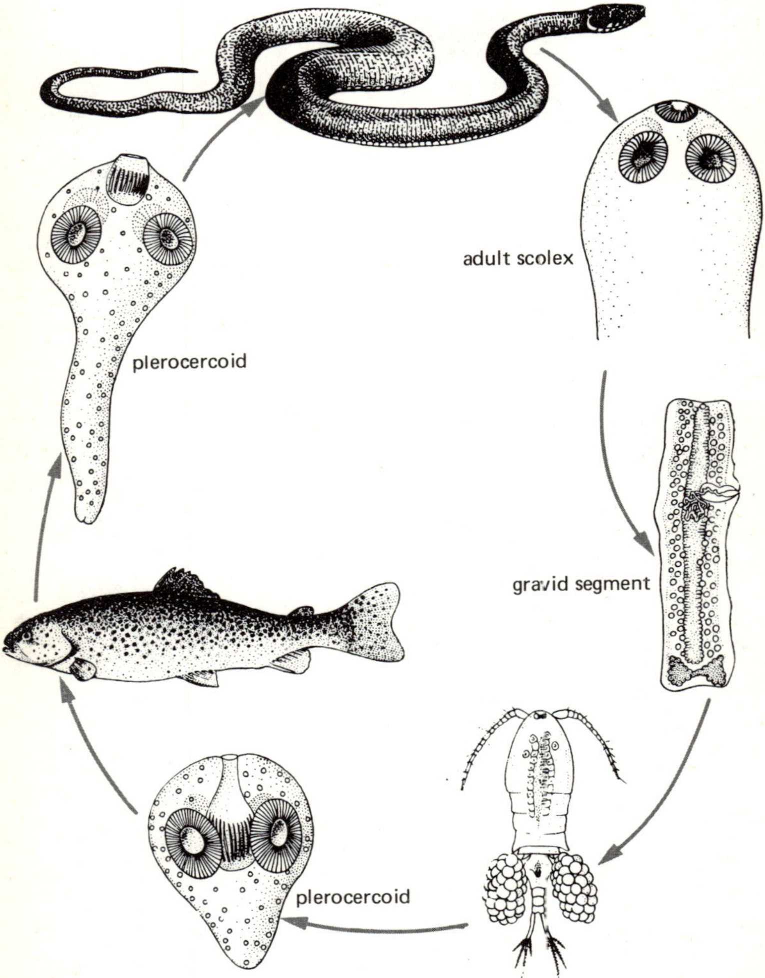

the ground, merely covering it with stones. Scavenging hyaenas find the bodies and in devouring them acquire their cestodes. Unlike *D. mansoni*, man is not a paratenic host here but a normal intermediate host.

Bothriocephalus scorpii occurs in the intestine of many marine fishes, especially the turbot, and is the only marine tapeworm whose life cycle is known. The first part of the life history up to ingestion of the ciliated coracidia by a marine copepod is similar to the life cycles already described above. The second intermediate host is a small fish, the larval cestode transforming into an infestive plerocercoid in its stomach. Unlike the preceding examples, this larva remains in the stomach where it is attached to the mucosa by the scolex. Segmentation begins and the genital rudiments appear, but then development ceases at this stage until such time as this fish is eaten by a larger fish when development recommences.

Ligula avium inhabits the intestine of aquatic birds, especially ducks. This species is characterised by the fact that the strobila does not grow in the definitive host and very much resembles a plerocercoid. The life cycle occurs in fresh water, the first intermediate host is again a copepod and the second a fish which harbours plerocercoids in the body cavity. These can reach a size of 25 cm and contain rudimentary reproductive organs. The bellies of the infested fish are often distended by the plerocercoids, and being overburdened in this way the fish tend to be more easily captured by the definitive host. Another peculiarity of this life history is the extremely short time spent by the adult worm in the duck intestine. The genital rudiments present in the plerocercoid mature four days after infestation of the duck and the whole worm is eliminated after twelve days, when the proglottids have produced eggs.

The life cycles of elasmobranch tapeworms should probably be dealt with here, though not one of them has been completely elucidated. It is known that the coracidium is eaten by a marine copepod and develops here into a procercoid, but so far the experimental infestation of a second intermediate host has not been accomplished. When plerocercoid larvae, recognisable by their scolices, are found in various marine vertebrate and invertebrate hosts it is not possible to decide whether they are present in a paratenic host or in the second intermediate host. In the case of a species of *Phyllobothrium* which is not uncommon in the fatty tissue of dolphins, it can be concluded that this is in a paratenic host since a shark tapeworm which utilised a cetacean as an intermediate host would stand little chance of completing its life cycle.

(*ii*) *With one intermediate host and a facultative paratenic host:* Species of the genus *Ichthyotaenia* occur in fresh-water fish and have a life cycle involving a single intermediate host and a paratenic host. The eggs of *I. pinguis*, a parasite of the American pike, are eaten by a copepod and the oncosphere larvae penetrate into the body cavity. In this first intermediate host a procercoid develops which will transform into a plerocercoid. Percids can also become infested and serve as paratenic hosts. Young pike, which are plankton feeders, can therefore acquire the infestation by eating infested copepods whilst the older carnivorous pike become infested by eating the paratenic percid hosts. Thus infestation seems to be doubly ensured in both young and mature pike. The paratenic host obviously remains facultative in plankton-eating hosts.

Ophiotaenia racemosa belongs to the same group of tapeworms but has a grass snake as its definitive host. The life cycle has been established experimentally and differs little from the course of events described above, save that the paratenic host here is a small fish. It is not impossible that the grass snake, living near water, could acquire infestation directly by eating infested copepods, but it

5·16 *Caryophylleus laticeps,* a neotenic cestode. (**a**) Adult worm; (**b**) procercoid larva developing in the body cavity of *Tubifex.*

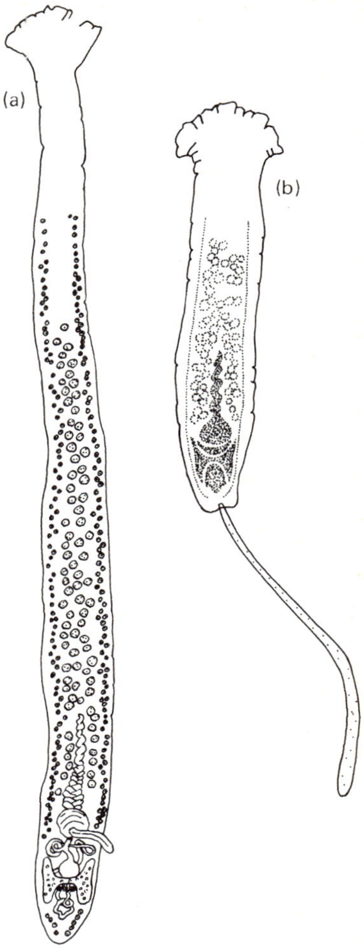

seems more likely that the snake would feed on small fishes and that although facultative from the point of view of parasite development, the paratenic host would in fact become quasi-obligatory to the life cycle.

Another unusual life cycle occurs in three species of hymenolepid living in ducks; *Hymenolepis collaris*, *H. coronula* and *H. gracilis*. Here the cysticercoid larvae are harboured by copepods and ostracods. Obviously weighed down by the larvae they contain, these entomostracans fall to the bottom or come into shallower water. Here they are eaten by and accumulate in aquatic molluscs, for the cysticercoids survive in the stomach of the limnaeids. In this manner a duck eating an infested mollusc is likely to acquire a multiple infestation. Thus the mollusc can be best considered as a reservoir host which accumulates larvae rather than merely as a paratenic host.

(*iii*) *With a single intermediate host:* Many species of fresh-water cyprinid fish have been found to contain somewhat unusual cestodes which have only a single set of male and female genitalia and are completely unsegmented (i.e. are monozoic). This is comparable to a cestode reduced to only the scolex and one proglottid and lacking a growth zone. All *caryophylleids*, a group containing many genera and species, have this kind of structure and it seems that here one is dealing with a neotenous plerocercoid, because the detailed anatomy of caryophylleids resembles that of a single proglottid of *Diphyllobothrium*. In addition the nature of the life cycle supports this hypothesis of neoteny. Eggs enclosing the oncosphere are eaten by small fresh-water oligochaetes of the genus *Tubifex*, where they develop into procercoids in the body cavity. These are unusual in that rudiments of the reproductive system are already in evidence and because a cercomere is present which bears the six larval hooks (figure 5·16). As soon as the *Tubifex* carrying the procercoids is eaten by a carp the procercoid loses the cercomere, grows and becomes mature, the adult worm really has the status of plerocercoid here.

Many species of tapeworms from marsh and water birds and

aquatic mammals use fresh-water invertebrates such as crustaceans, insects and molluscs as hosts. The woodcock, for instance, harbours a tapeworm, *Paricterotaenia embryo*, whose eggs are eaten by the leech, *Herpobdella atomaria;* the cysticercoids develop in the tissues of the leech. Also the water shrew often carries several species of *Hymenolepis* which use gammarids as the intermediate host.

(*iv*) *With no intermediate host:* Only very few direct life cycles are known for cestodes and of the few described some are spurious due to incomplete observations or faulty experiments. In the body cavity of *Tubifex*, however, occurs the only adult tapeworm found in an invertebrate host, and for this a new genus, *Archigetes*, had to be created. Liberation of the eggs occurs only when the host dies. These are then eaten directly by another *Tubifex* and the oncospheres penetrate into the coelomic cavity where they develop into adult worms which are procercoid-like and unstrobilated rather like the caryophylleids. It seems likely, therefore, that *Archigetes* is really a neotenous procercoid larva, which explains the presence of a mature cestode in the body cavity of an invertebrate definitive host. This worm is the only example of its genus.

(b) Terrestrial cycles:

(*i*) *With two hosts, one facultative:* The genus *Mesocestoides* is a parasite of carnivorous bird and mammal hosts and its life cycle resembles that of *Diphyllobothrium* in that there are two intermediate hosts, the second of which is probably paratenic. The eggs are eaten by oribatid mites in which a special kind of larva develops which resembles a plerocercoid but has an invaginated scolex bearing four suckers; this is called a *tetrathyridium*. Such larvae have been found to occur in the body cavity of a large number of hosts, even carnivores, and in most cases it is assumed that these are acting as paratenic hosts rather than as obligatory intermediate

5·17 Life cycle of *Hymenolepis integra* (aquatic cycle).

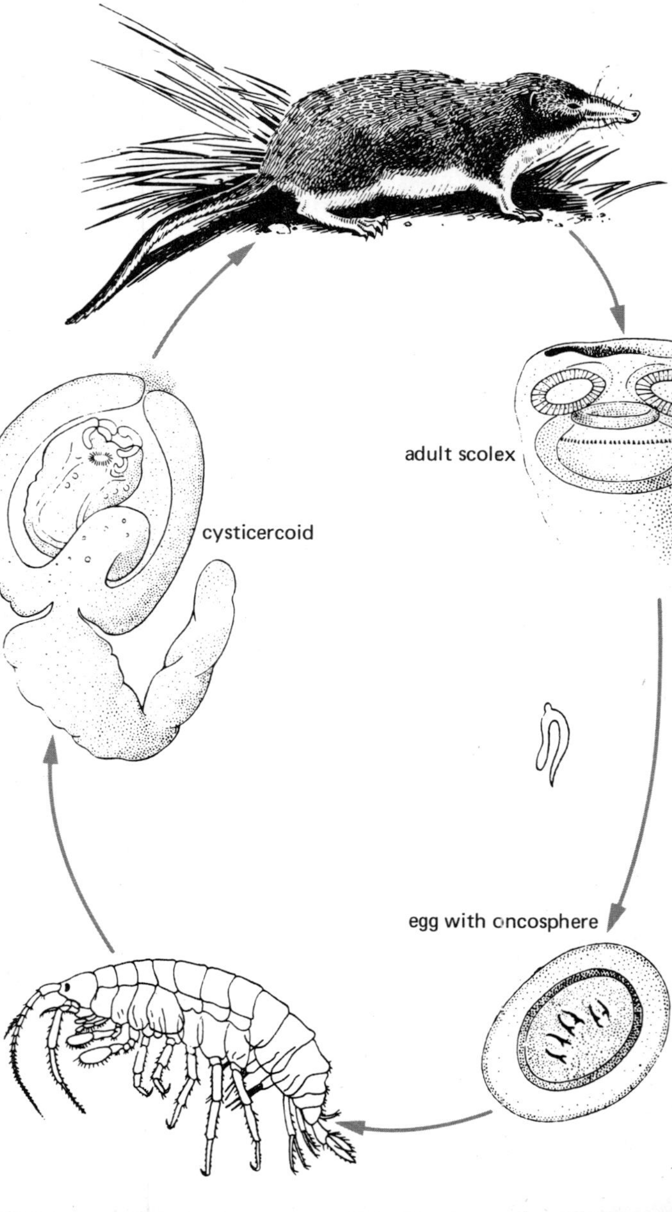

5·18 Life cycle of *Raillietina bonini* (terrestrial life cycle).

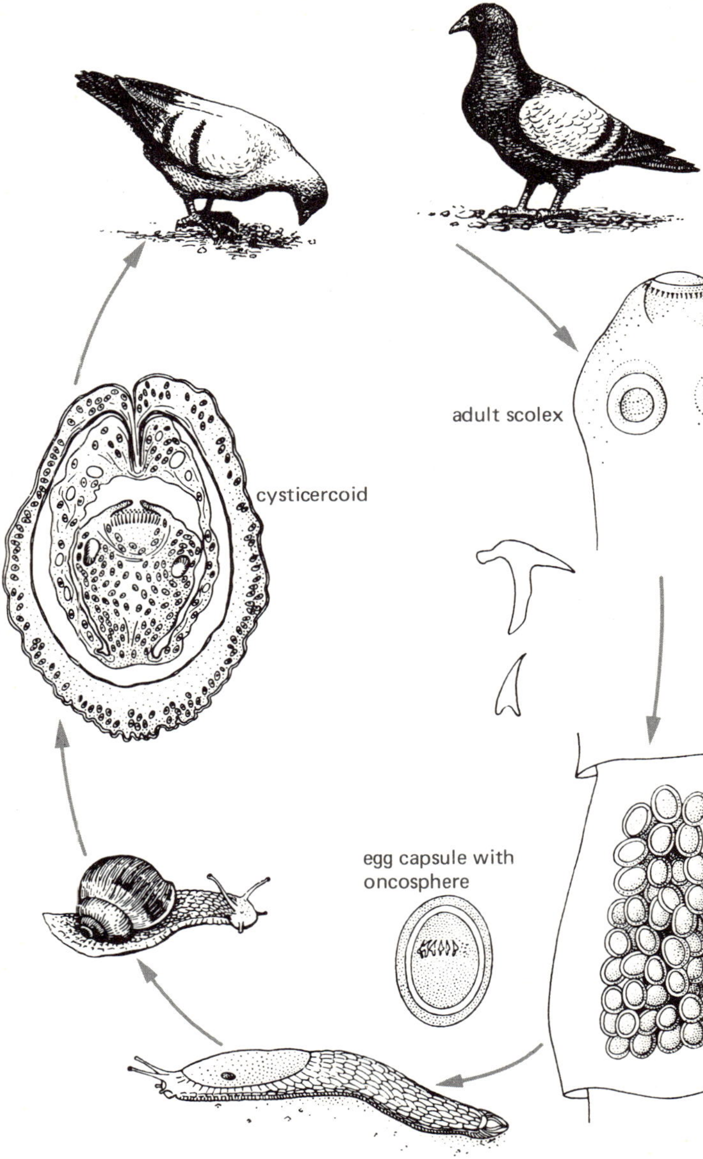

hosts. Nevertheless it seems more likely that a fox would eat a small mammal carrying larvae rather than a mite and it is possible that, in this case, the paratenic host is obligatory.

(*ii*) *With a single intermediate host:* This kind of cycle is characteristic of cestodes parasitising vertebrate hosts and involves sometimes vertebrate, sometimes invertebrate intermediate hosts. The nature of the larva is adapted to the type of host in question and is a cysticercus in vertebrate hosts and a cysticercoid in invertebrates. Larval multiplication occurs in some forms and an enormous number of larvae can be produced from a single egg, resulting in a massive infestation of the intermediate host.

Dipylidium caninum is a common parasite of the dog and cat. The eggs are voided in capsules, each containing about fifteen eggs. The intermediate host is the dog flea. Adult fleas have fine tubular mouthparts adapted for blood sucking so they are incapable of taking in cestode eggs. It is, in fact, the flea larvae living on the detritus of the kennel or cat basket which become infected. These larvae have crushing mouthparts used for eating small fungi and micro-organisms and are attracted to the isolated cestode proglottids which they break open. Eggs swallowed by these larvae hatch liberating oncospheres which penetrate the wall of the gut and enter the body cavity. These immediately start to develop into cysticercoids, but their presence triggers off a host defence reaction and the cestode larvae become surrounded by leucocytes which inhibit further development. Just before the metamorphosis of the flea, however, these leucocytes are mobilised in order to phagocytose and destroy larval tissues preceding formation of the pupa. Freed from restraint the cysticercoids complete their development so that the adult fleas emerge containing infestive cysticercoids.

Domestic pigeons often harbour *Raillietina bonini* and eggs are voided in the droppings. The intermediate host is a mollusc, either a slug or a snail, which eats the eggs and which contain cysticercoids. Humidity is necessary to completion of the life cycle, dry conditions inhibiting development.

Catenotaenia pusilla occurs in young white mice and wild rodents and uses a detritus feeding mite of the genus *Glyciphagus* as intermediate host. The cysticercoid of this tapeworm is unusual in that it lacks the adult scolex with four suckers and has instead a single large apical sucker which is used to attach the larva to the mucosa while the four suckers differentiate and become functional. The apical sucker is then gradually resorbed.

Larval multiplication in the invertebrate intermediate host is particularly a feature of the genus *Hymenolepis*, which occurs mainly in birds but also in some mammals.

In *Hymenolepis cantaniana*, a parasite of turkeys and pheasants, a dung beetle devours the parasite eggs and the cysticercoid develops a branching form which buds off many scolices, thus ensuring from the outset infestation with many worms.

The white-toothed shrew harbours a very small tapeworm, *Hymenolepis pistillum*, which is always present in large numbers. It develops in the myriapod *Glomeris*, inside which the oncosphere transforms into a germinal mass which buds off scolices; each tissue mass can produce fifteen to twenty scolices. A single *Glomeris* may eat several eggs, in which case they become packed with scolices so that a massive infestation can be acquired by the shrew eating the intermediate host.

The common shrew often carries a large number of minute tapeworms, *Hymenolepis prolifer*, which consists of only eight proglottids. The larval forms of this species also occur in myriapods where larval multiplication takes place. Unlike the former case the scolices detach from the germinal mass as soon as they are fully formed and fill the body cavity of the myriapod. Up to a thousand can occur in a single intermediate host.

5·19 Life cycle of *Hymenolepis pistillum* (proliferative larva).

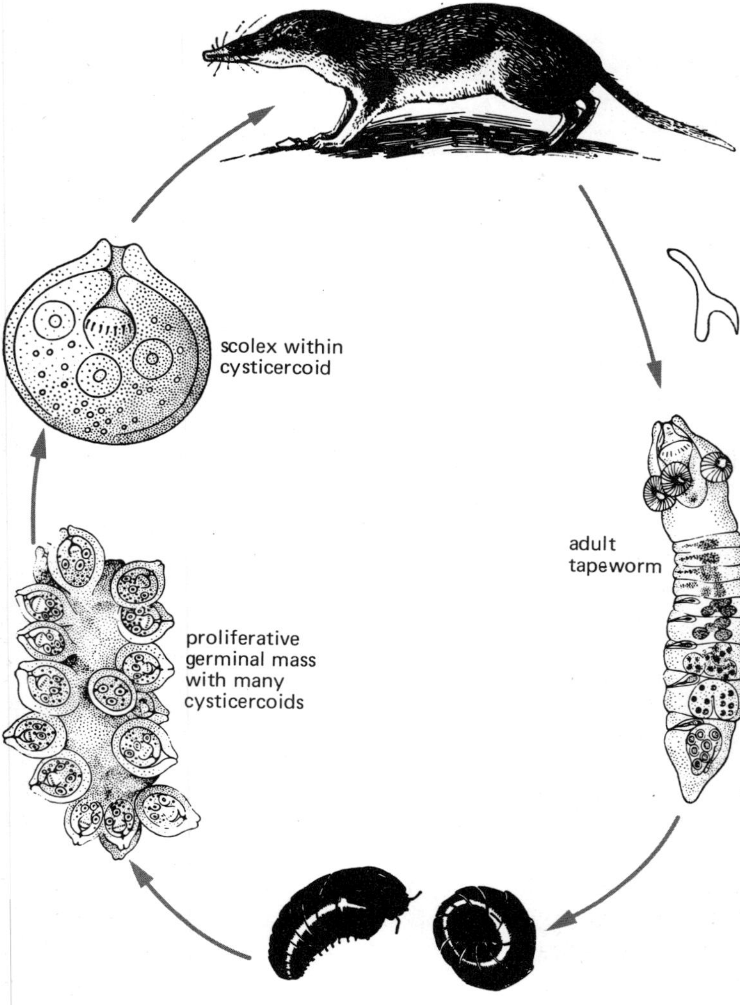

5·20 Proliferative larva of *Hymenolepis pistillum* in *Glomeris* (myriapod).

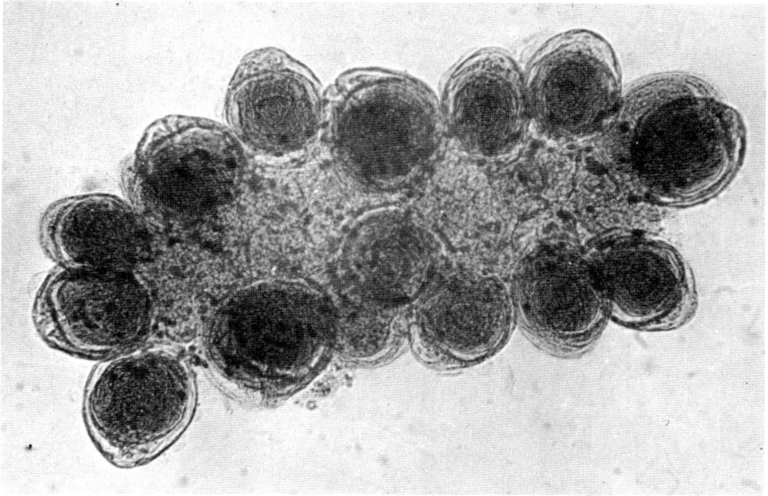

5·21 Below Proliferative larva of *Hymenolepis prolifer* in *Glomeris* (myriapod). **Left** Proliferative embryonic mass. **Right** Mature detached cysticercoid showing the crown of hooks and the suckers.

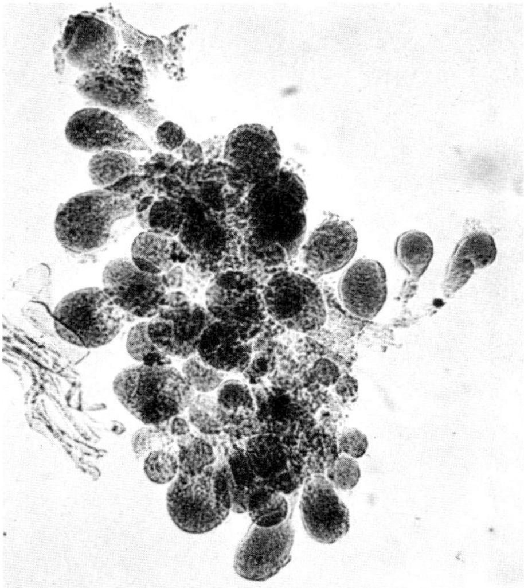

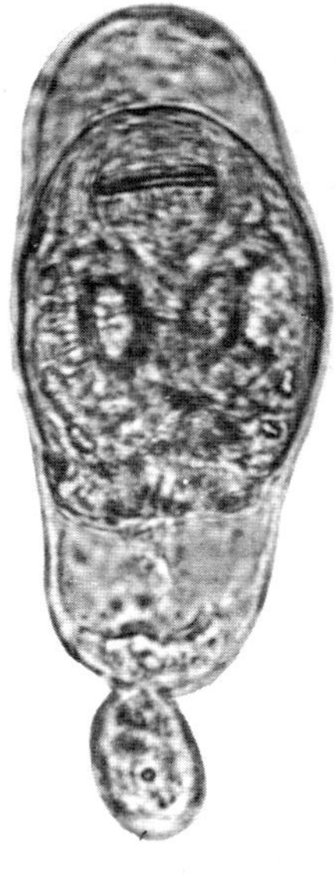

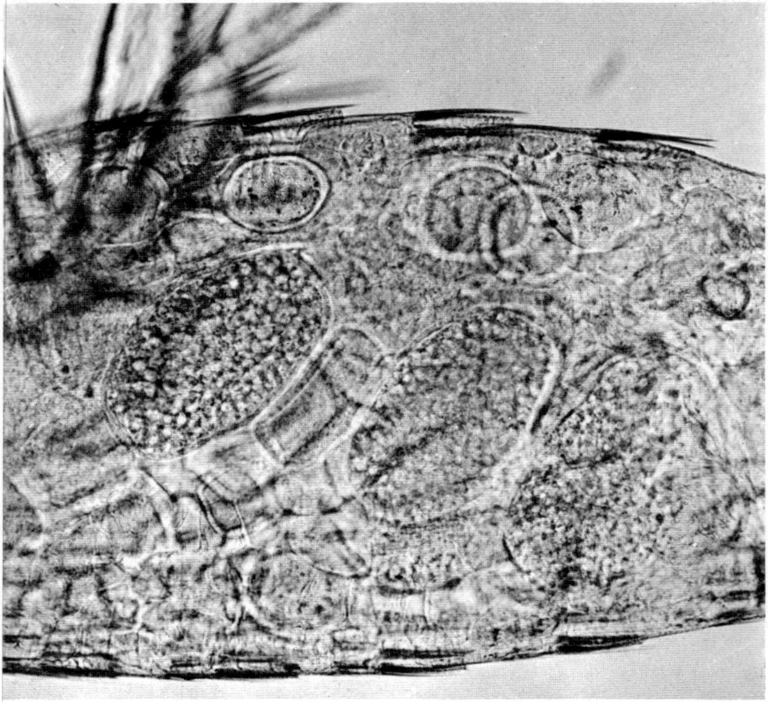

Cysticerci can always be seen with the naked eye and in general characterise the life histories of taeniids, which are parasites of man and carnivores. In *Taenia solium*, for example, a form specific to man, the eggs are voided within the proglottids, and in conditions of primitive sanitation will often get spread around on the ground. Pigs encounter and swallow the eggs and the oncospheres penetrate the intestinal mucosa, enter the circulatory system and are carried to all parts of the body in the blood. The cysticerci grow slowly until they are about the size of a pea. At this stage the scolex is completely invaginated into the anterior region but already bears a double ring of hooks of characteristic shape and size. If several cysticerci are swallowed only one usually develops into an adult, and this is probably due to competition for nutrients, because *T. solium* is a large worm, between 5 to 7

5·22 Opposite Cysticercoids of *Hymenolepis microstoma* in the body cavity of a mouse flea.

5·23 Below An X-ray picture of the pelvic region of a man suffering from cysticercosis produced by the cysticerci of *Taenia solium*.

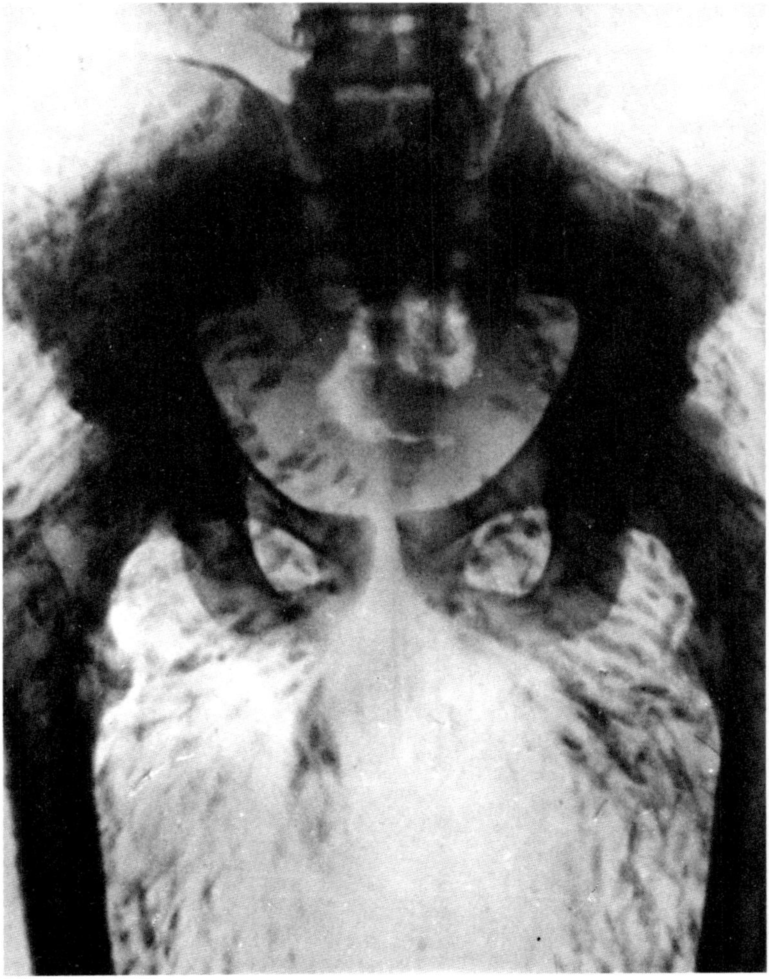

5·24 Top Section through the gall bladder and intestine of a white mouse. *Hymenolepis microstoma* occurs in the gall bladder and *Catenotaenia pusilla* in the intestine.
5·25 Bottom Polycephalous cysticercus of *Taenia parva* from a field mouse from the south of France.

metres long and about 5 mm wide. The pig is the usual intermediate host but can be replaced by the wild boar; man himself can also harbour cysticerci and sometimes in large numbers. Human cysticercosis, caused by man swallowing the eggs, is a real medical problem in certain regions of tropical South America and Africa. Except where cannibalism occurs, human cysticercosis represents a blind alley for the life cycle, since man is the only known definitive host for this species.

Dogs of some regions harbour *Taenia pisiformis*, the larvae of which develop in rabbits. It is not surprising that there is a high incidence of infestation in hunting dogs.

Oncospheres hatching in the intestine of the rabbit burrow through the gut wall and enter the venous capillaries and from here pass via the portal veins into the liver. In the liver growth of the cysticercus commences and this reaches a size of about 10 mm before quitting the liver and boring out through the connective tissue capsule of this organ to continue its development in the mesentery. Here the cysticercus reaches the size of a large pea. The mature cysticercus has the same structure as that described for *T. solium*. This is why when huntsmen run a rabbit to earth, dogs are not permitted to eat the entrails.

Many cats harbour a very large tapeworm, *Taenia taeniaeformis*, the scolex of which is visible to the naked eye. The larvae develop in the liver of various rodents – mice, field mice, voles – and can be seen, under the connective tissue capsule surrounding the liver, as round white cysts. The fully formed cysticercus is extremely large, measuring 3–40 cm, giving the impression that strobilisation may already have commenced in the cyst. In fact a large part of this is a hollow larval structure which becomes separated from the scolex in the intestine of the cat about five hours after infestation has occurred.

In *Taenia crassiceps*, which occurs in foxes, several worms are always found together, which indicates that larval multiplication

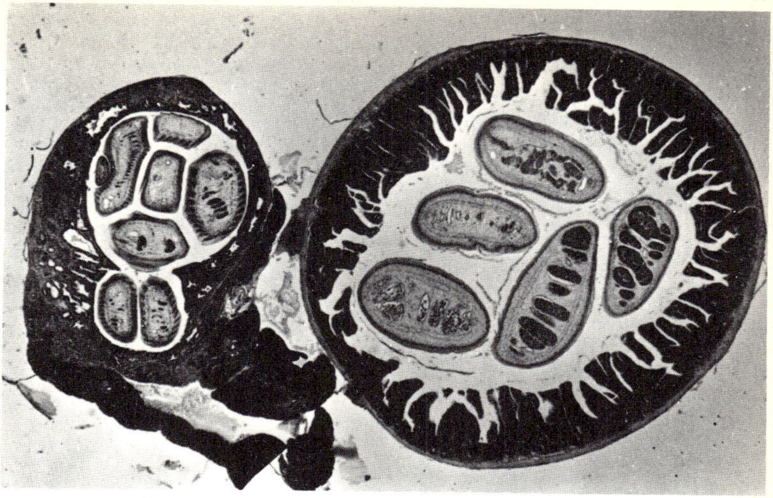

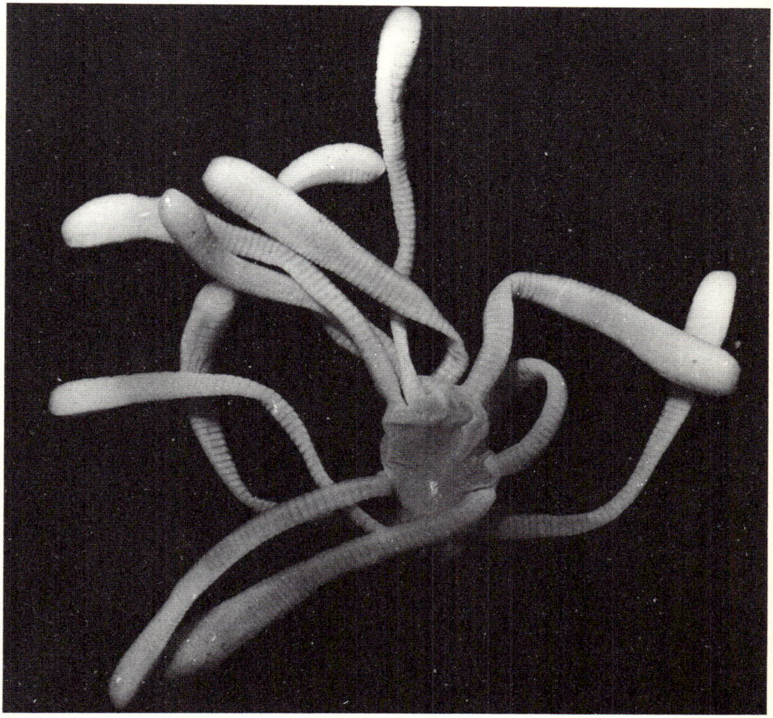

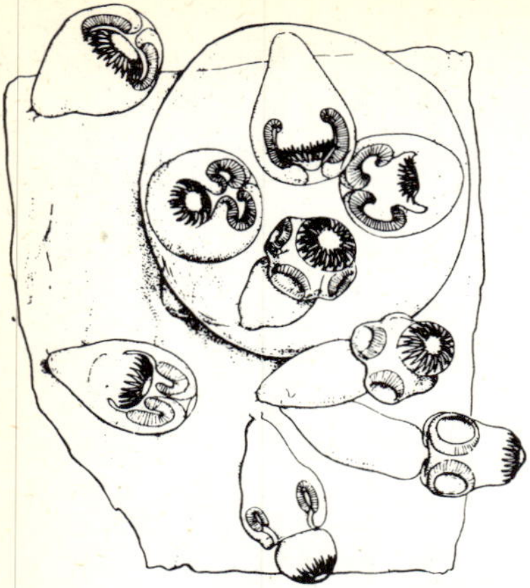

5·26 Left Fragment of the germinal membrane of the hydatid cyst of *Echinococcus echinococcus.*

5·27 Below Section through the liver of a vole parasitised by *Echinococcus multilocularis.*

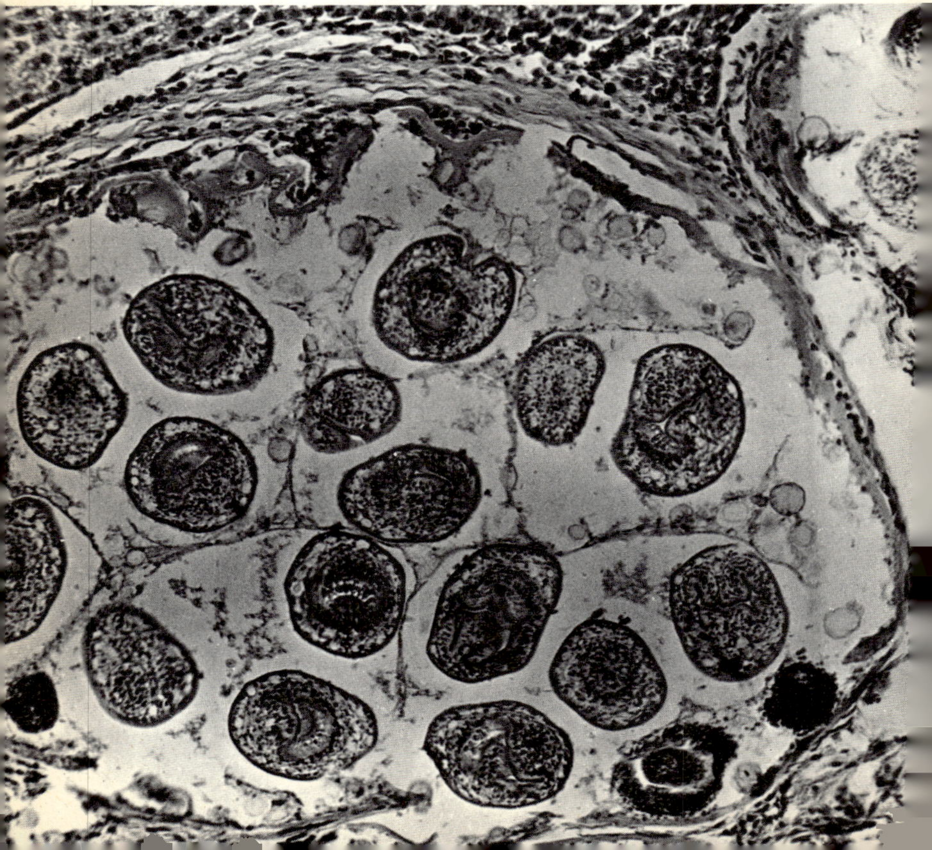

probably occurs. In fact, in the intermediate rodent host the cysticerci form huge masses, often beneath the skin but also in the abdominal cavity, where hundreds of larvae are tightly packed together. When a detailed examination of these larvae is made, after separating them out in physiological saline, it can be seen that while some of them resemble elongated cysticerci with an invaginated scolex, other less dense larvae bear a proliferating bladderlike region at the posterior end of the body and show no trace of a scolex. These detach and give rise to new buds. Each of the latter acquires a scolex and becomes an infestive larva. In this case it is not the scolices which bud but the larval 'bladder', the scolex being developed subsequently. The proliferative bladders can continue to multiply even if they fall into the peritoneal cavity.

A secondary cysticercus formed in this way can be maintained artificially by subculturing. We have been able to keep one for eight years in this manner.

The genet of southern France, Spain and Africa harbours *Taenia parva* which again occurs in numbers. Although the life cycle has not been followed experimentally, it is possible to identify the larva from the number, size and shape of the hooks. The larva occurs in the pleural cavity of rodents and is a multicephalate cysticercus containing about twenty stalked scolices in a common cavity (figure 5·25).

Taenia serialis is another parasite of the dog, and the cysticercus larva of this worm always encloses a large number of scolices arranged in longitudinal series in the cavity of the bladder, giving this species its name. This larva occurs in the mesentery of rabbits and can grow to a diameter of several centimetres.

The smallest tapeworm known is *Echinococcus echinococcus*, which is merely 6 mm in length and is made up of only three or four segments; it has the largest larva, however, which produces thousands of larval scolices. This can develop in nearly all groups of mammals, including man. Human echinococcosis is known medically as *hydatid cyst*. Man is obviously not a very suitable intermediate host from the point of view of furthering the life cycle. From an epidemiological point of view hydatidosis is of considerable importance and has resulted in much research being devoted to this curious larval stage and its development. Most organs of the intermediate host are susceptible to implantation of hydatid cysts, although the lungs and liver are most heavily parasitised. Dogs obtain the infestation from contact with farm stock such as cattle and sheep, so that echinococcosis tends to be common in countries where these animals are reared.

The hydatid cyst develops very slowly; after five months in the liver the larval cyst measures only 1 cm in diameter, and it takes several years before a size of 10 to 15 cm is reached. The lining of the cyst which produces the scolices is termed the *germinal epithelium*. By a special budding process some scolices develop into *daughter cysts* which in turn bud off a new generation of scolices from their inner wall. A large hydatid cyst may therefore contain many daughter cysts as well as thousands of scolices, so that when a dog becomes infested hundreds of small cestodes are acquired simultaneously. A special feature of the hydatid cyst is that like the cysticercus of *Taenia crassiceps* already mentioned, if this is ruptured secondary echinococcosis results. If this occurs, each scolex liberated into the peritoneal cavity is capable of producing a new daughter cyst. It is easy to see why general hydatidosis of man is always quite a serious condition.

(*iii*) *With one facultative intermediate host:* It might seem that a life cycle in which the intermediate host is facultative is something of a paradox, since in this case the life cycle would be direct. This does in fact at first sight appear to be the case; mice for instance, can be infested by feeding them mature eggs of *Hymenolepis fraterna*.

5·28 Life cycles of *Hymenolepis fraterna*.

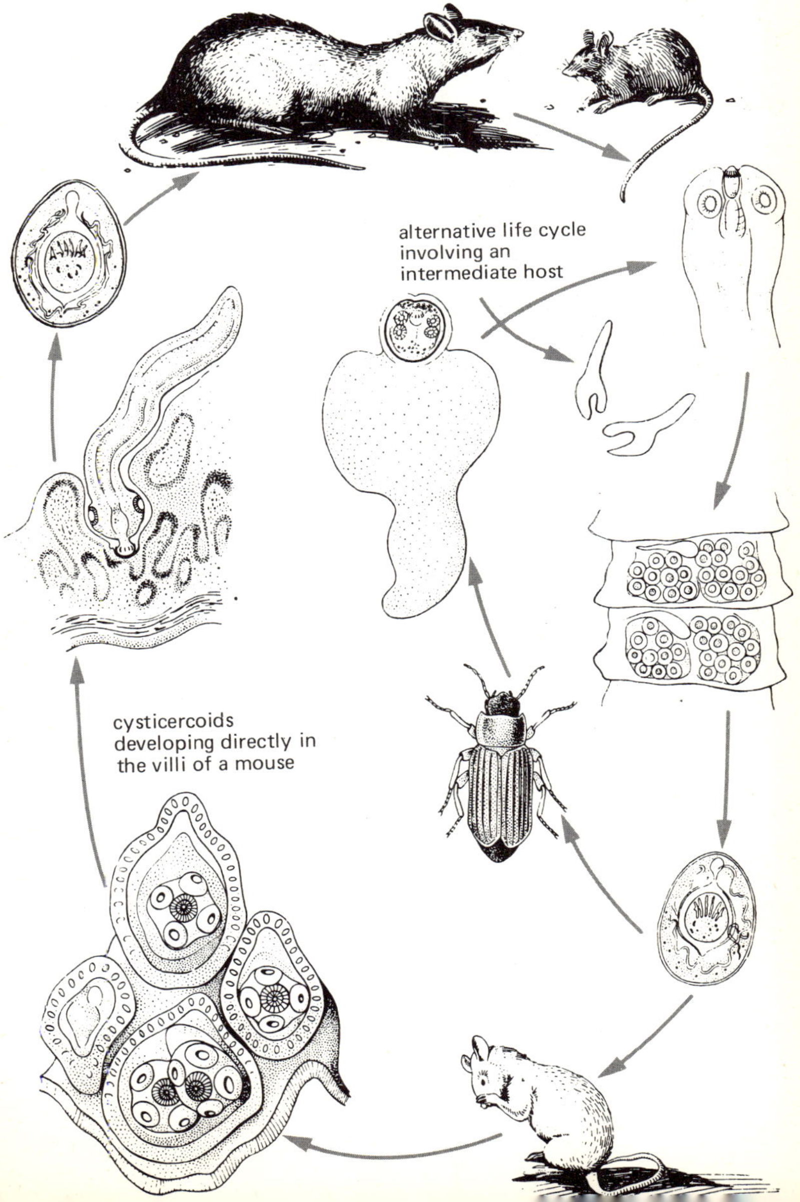

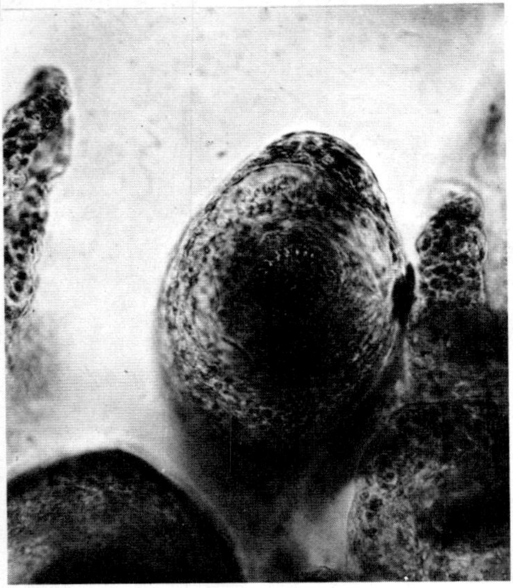

5·29 Preparation of an intestinal villus of a mouse containing a cysticercoid of *Hymenolepis fraterna*.

After several days they can be found to contain mature egg-producing adult worms. It the experiment is repeated, using several mice and the first killed about three days after infestation, one finds cysticercoids situated within the villi of the first part of the intestine. A mouse killed about six days after infestation no longer harbours cysticercoids within the villi but contains young worms attached by their scolices to the posterior part of the intestine. By following the infestation throughout the various stages it can be seen that the oncosphere larvae liberated in the stomach of the mouse then penetrate into the intestinal villi and develop into cysticercoids in three to four days. After this the larvae break free and attach to the intestinal mucosa. The presence of a cysticercoid larva in a vertebrate is quite exceptional, but this is due to the modification of a life cycle involving a single intermediate host. The alternative cycle involving an intermediate host, which is an insect, is probably the primitive condition. When such hosts take in eggs of *H. fraterna*, cysticercoids develop in the haemocoel, but these show certain superficial differences from those occurring within the intestinal villi of the mouse. The scolices of both forms are identical but the bodies of the larvae, which have a trophic function, are very different, being much larger within the insect than within the mouse villus (figure 5·28). It seems not unlikely that the surface area of the larva available for uptake is proportional to the amount and type of nutrient present in each situation. In the apparently direct life cycle of this tapeworm, the intestinal villi of the mouse act as the intermediate host so that both here and in the case of its *nana* form, which is adapted to man, definitive and intermediate host are the same animal.

Hymenolepis grisea from European bats may have a similar kind of life cycle but the experimental evidence is insufficient to decide this matter; nor is the intermediate host known, though this may be a guanophilic insect.

This brief review of the basic kinds of life cycle, although incomplete, does illustrate, both in the case of aquatic and terrestrial life cycles, a tendency towards reduction of the number of intermediate hosts from three to one. The total loss of such hosts is connected with neoteny in aquatic cycles and with incorporation of intermediate and definitive hosts into the same individual in the terrestrial forms. Two different kinds of evolutionary process are thus involved. The role of the paratenic host and the ecological consequences of using different kinds of intermediate host on the distribution of cestodes in vertebrates will be dealt with later.

Trematodes

Trematodes are unsegmented flat worms which have a medianly situated ventral sucker as their main attachment organ. This organ can, however, be lost in some forms. The mouth, often surrounded by an oral sucker, leads towards a muscular pharynx used for pumping liquid food material into the gut. This is usually made up of two caeca which may occasionally cross-connect and which

5.30 Diagram of the life cycles of trematodes. For a full explanation see the text.

may in extremely rare cases open by a terminal anus. There is some evidence to suggest that, despite the presence of a gut, trematodes can absorb nutrients over the external cytoplasmic epidermis, although this is not such a complex structure as the cestode epidermis and lacks microtriches. The excretory system opens posteriorly by means of a bladder which is fed by two main lateral ducts, and these in turn receive small canals terminating in protonephridia. The arrangement of the excretory system is of considerable importance in digeneans because, in addition to being a useful taxonomic feature in general, it allows one to distinguish between the different cercaria larvae.

As a rule trematodes are hermaphrodites but there is a secondary tendency towards separation of the sexes and sexual dimorphism. Adult trematodes live in natural cavities of vertebrates, the intestine, bile and pancreatic ducts, bladder, lungs, cranial sinus, as well as in mucosal cysts. Wherever the worm lives the eggs are voided via natural passages.

The life cycle is heteroxenous and is characterised by larval multiplication or polyembryony in the first intermediate host, a mollusc. This culminates in the production of cercariae which usually then leave the first intermediate host. A series of larval stages occurs within this first mollusc host, starting with the *miracidium* larva formed within the egg. This enters the tissues of the mollusc either by active penetration or passive ingestion within the egg shell and often ends up in the digestive gland, which is rich in food reserves. Here the miracidium elongates and transforms into a kind of larva, the *sporocyst*, which is sac-like and produces daughter sporocysts and rediae by division of germ cells present in the miracidium. Unlike sporocysts, the rediae possess a mouth, followed by a pharynx which opens into a short straight intestine (figure 5·35). The daughter sporocysts are usually smaller and more deeply pigmented than the mother sporocysts but both give rise

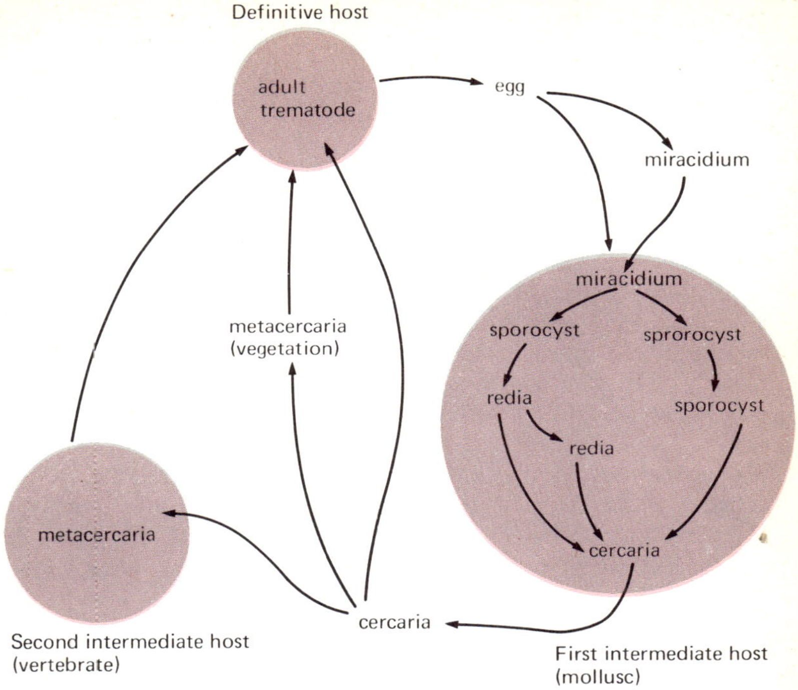

ultimately to cercariae. The cercariae usually represent the final stage of development occurring in the first intermediate host. They resemble small adults for the suckers and gut are already developed to differing extents and there is an excretory system with the form characteristic of the particular adult worm. Two important features of the cercariae are the propulsive tail, which may be simple or bifid, and the glands responsible for penetration into the second intermediate host, also those associated with encystment. The penetration glands open in the region of the oral sucker and may work in conjunction with a mobile stylet, while the cystogenous glands are diffusely arranged at the margins of the body. Cercariae are usually free-living for a short while after leaving the mollusc. There are then three possible alternatives: encystment in the ambient medium on vegetation or an inert object; penetration into the second intermediate host with subsequent encystment; or direct penetration into the definitive host. In the first two cases the cercaria encysts after losing its tail and resorbing the penetration and cystogenous glands. Considerable differentiation has occurred by this time and rudiments of the reproductive system are already apparent. This encysted stage is termed the *metacercaria* and represents the infestive larva which will transform into an adult trematode in the definitive host. Where direct penetration via the skin of the definitive host occurs, the cercaria enters the blood capillaries and develops directly into an adult worm without leaving the circulatory system, thus bypassing the encysted metacercarial stage. This is an exceptional kind of life cycle which is characteristic of blood flukes. It must be emphasised that the complex embryology of trematodes in molluscs contributes greatly to the efficiency of the life cycle by augmenting to some considerable degree the number of infestive larvae produced. Larval multiplication may continue within the mollusc for as long as this survives. A littorinid infested with redial stages of a trematode living in the herring gull as an adult, and isolated so as to prevent reinfestation, produced five and a half million cercariae in the first five years and after seven years 1,600 cercariae were still being produced daily. The experiment was terminated only because an accident occurred. In this particular case the second intermediate host is a herring, which is a shoaling fish occurring in large numbers. It is therefore possible that production of a correspondingly large number of cercariae may allow an even distribution of metacercariae throughout the fish population, thus eventually ensuring infestation of a large number of gulls.

In the following description of the comparative biology of trematode life cycles it must be emphasised that these life cycles, though following a basic plan, have tended to become adapted to particular ecological conditions and show various modifications. It must be pointed out that consideration of these life cycles in this way in no way reflects the relationships between different groups of adult worms or demonstrates evolutionary pathways within the group. Compression of the life cycle often occurs due to the phenomenon of *progenesis*, that is, precocious maturation and self-fertilisation of the metacercariae which produce viable eggs. Progenesis always occurs at the metacercarial stage and cannot, therefore, be confused with neoteny, although both result in abbreviation of the life cycle so that the definitive host becomes facultative or is made redundant. The distinction made for the cestodes between aquatic and terrestrial life cycles is not practicable here, since it is often difficult to separate the two, but mainly because this does not reflect evolutionary tendencies in the form of the life cycle as it does in the cestodes.

(a) Life cycles with a single intermediate host where the cercariae remain within the intermediate host:

(*i*) *Cercariae encyst within the sporocyst: Plagioporus sinitsini* parasitises the gall bladder of a fresh-water fish and its miracidia penetrate into a mollusc and develop into a primary sporocyst which produces vividly coloured yellow or red daughter sporocysts. The cercariae formed within the sporocysts have only a stub of tail (cotylocercs) and encyst within the daughter sporocysts. Remarkably enough the coloured sporocysts, which also contain metacercariae, are eliminated via the gut of the mollusc and can survive for at least twenty-four hours outside the intermediate host. Attracted by those small, coloured larvae, the fish eats them and becomes infested. Reduction of this life cycle is probably secondary because other species of the genus *Plagioporus*, which are also parasites of fresh-water fish, have a more normal life cycle involving two intermediate hosts.

A similar kind of life cycle occurs in *Diphterostomum brusinae*, which is a parasite of marine fish, but this differs slightly in that the

5·31 The snail (*Succinea putris*) with tentacles parasitised by the sporocysts of the trematode *Leucochloridium paradoxum*, which contains metacercariae. (1) Normal tentacles; (2) both tentacles enclose brown-banded sporocysts; (3) one tentacle contains a brown-banded and the other tentacle a green-banded sporocyst; (4) a sporocyst in process of being discharged; (5) the ruptured sporocyst liberating metacercariae; (6) the adult worm packed with eggs.

sporocysts do not leave the mollusc and this has to be eaten by the final fish host. That this trematode occurs only in mollusc-eating fish is, then, hardly surprising.

The life cycle of the genus *Leucochloridium* is especially adapted for infestation of many kinds of passerine definitive hosts. The molluscan first intermediate host is always a marsh-living *Succinea* species; this is a group of molluscs which lives on floating vegetation or on plants growing at the edges of streams. The miracidium develops within the mollusc into a much-branched sporocyst within which tailless cercariae are formed. These encyst within the sporocyst and each metacercaria can be seen to be surrounded by a thick protein coat. The metacercaria-filled sporocysts become very large, up to 12 mm in length, and pulsate rhythmically; their walls are striped with alternating red and green or brown bands, giving them the appearance of small caterpillars.

At this stage the sporocysts penetrate a tentacle of the mollusc which they dilate and cause to pulsate. When the mollusc moves on to a leaf the movements of the tentacles and their pronounced colour attract birds. The tip of the tentacle usually only has to be touched, either by the leaf or by the bird's beak, for the sporocyst to break out and creep around on the leaf, where it is eaten by the bird. A single parasitised mollusc can product sporocysts containing metacercariae for a considerable time and can thus be a source of infestation for many birds.

(*ii*) *Cercariae encyst within the mollusc:* The life cycle of *Typhlocoelium cymbium* is fairly specialised because rediae are formed precociously. Adult worms inhabit the tracheae and nasal passages of many water birds and the eggs they contain hatch in the uterus. The miracidium already contains a redia and has to penetrate into a planorbid snail. In the latter the redia is released and produces

1

2

3

4

5

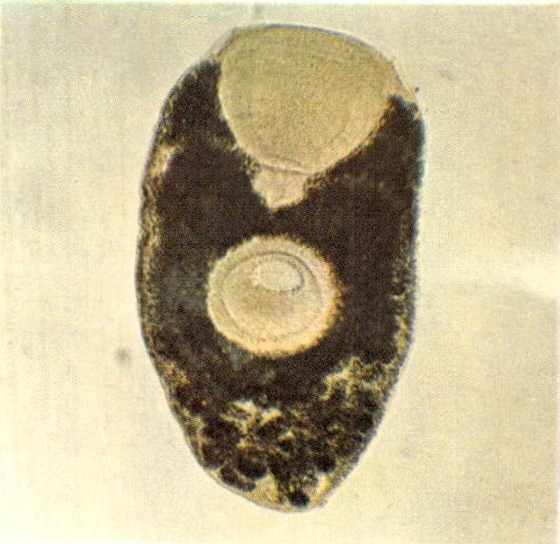
6

only two cercariae at a time, which lack a tail. The cercariae escape via the birth pore of the redia and encyst in the surrounding tissues of the mollusc.

(b) Life cycles with a single intermediate host where the cercariae leave the intermediate host:

Another unusual life cycle occurs in those parasites of fresh-water fish belonging to the family Azygiidae. The eggs are eaten by a mollusc and the miracidium hatches in the intestine of this intermediate host. It transforms into a sporocyst from which rediae escape; these grow and often accumulate in the pallial cavity of the mollusc. The intestine of these rediae degenerates and the mouth comes to serve as a birth pore for cercariae which are formed only two at a time. These cercariae have a massive forked tail into which they can completely retract themselves. Penetration glands and cystogenous glands are lacking in this extremely large cercaria, which reaches a length of 27 mm. The cercaria is liberated into the water where it floats head downwards, supported by the caudal furcae which adhere to the water surface. It soon attracts a fish by the movements of its body, which resemble those of a mosquito larva, and is eaten. Thus a metacercarial stage is absent from the cycle. In some species the cercariae have become neotenous, so it would be a small step from this to a situation where only one host, serving as both intermediate and definitive host, was involved.

As has already been mentioned, trematodes living in the blood system of their hosts have a special kind of life cycle. The eggs are usually large and pass through the tissues of the host until a natural cavity of some kind is reached from which they can escape to the exterior. The miracidium actively penetrates a mollusc and develops into a sporocyst, which in turn produces daughter sporocysts within which cercariae develop. The cercariae of all blood flukes have a forked tail (furcocercariae) and are equipped with penetration glands which are used to bore as far as the lymph vessels of the host. The schistosomes belong to this group of trematodes and these are characterised by sexual dimorphism, the female being lodged in the gynecophoric canal of the male so that permanent contact between the two sexes is maintained (figure 5·36). An interesting adaptation of this kind of cycle occurs in *Opisthioglyphe ranae*, which inhabits the intestine of frogs and newts. Larval development occurs in several species of limnaeids, and cercariae escaping from the sporocysts are equipped with a straight tail, penetration glands and a stylet which is situated where the ducts of the penetration glands become confluent (xiphidiocercariae). The cercariae leave the mollusc and penetrate into the branchial chamber of tadpoles, encysting in the buccal mucosa. When the tadpoles metamorphose the lining of the buccal cavity becomes altered; the metacercariae are rejected and then swallowed by the young frog, to become adult in the frog's intestine. It seems likely that this is an example of an originally two intermediate host life cycle which has become secondarily shortened since frogs can be fed metacercariae which have encysted, not in the mouth but in the skin of tadpoles, and these will develop into adult worms. Cannibalism seems to be a fairly unusual occurrence, however, so that a three-host life cycle is less likely to occur.

A life cycle can become abbreviated by the elimination of the definitive host, as for example when progenesis occurs in the metacercaria. This is true of *Ratzia parva*, which encysts in the skin of *Discoglossus*. Cercariae arising from the rediae lack a stylet but are equipped with penetration glands. They bore into the skin of *Discoglossus* and here transform into metacercariae. These become progenetic and lay eggs within the cysts, giving them a dark appearance. When egg laying has finished, the metacercariae die and are eliminated by histolysis of the cyst wall so that eggs are released into the water at the same time. Despite experimental

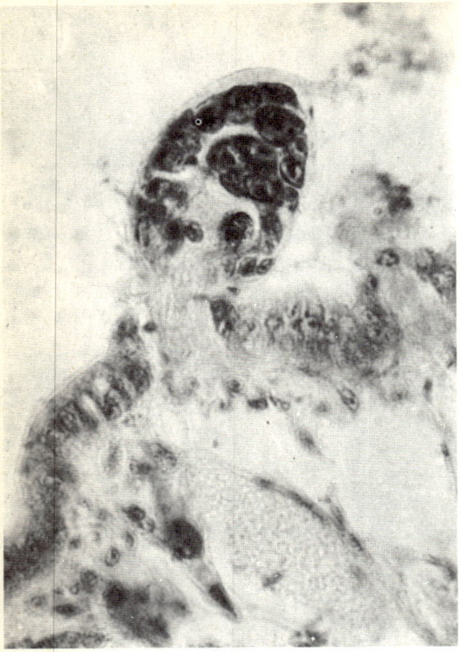

5·32 Left Penetration of the miracidium of *Fasciola hepatica* into the first intermediate host (notice the destruction of the epithelium of the mollusc).

5·33 Opposite Types of cercariae.
(**a**) Cercaria of *Bucephalus polymorphous*;
(**b**) macrocercous cercaria;
(**c**) trichocercous cercaria;
(**d**) furcocercaria;
(**e**) xiphidiocercaria;
(**f**) cystophore cercaria.

work and epidemiological studies, no host other than *Discoglossus* has been found for this fluke.

External encystment of the cercaria on aquatic plants or on the shell of the mollusc producing the cercariae occurs in many different groups of trematodes. The classic example is that of the liver fluke which inhabits the bile ducts of many mammals, mainly ruminants, but also on occasions man. This was the first trematode life cycle to be described and was discovered simultaneously in England and in Germany. Attempts are often made to relate all fluke life cycles to this one, but this would be an over-generalisation since the life cycles of fasciolids are specialised in that they lack a second intermediate host.

The miracidium penetrates into a limnaeid mollusc, which is not strictly aquatic but is a mud-living snail found near the edge of waterlogged ruts or small pools. Here it develops into a sporocyst

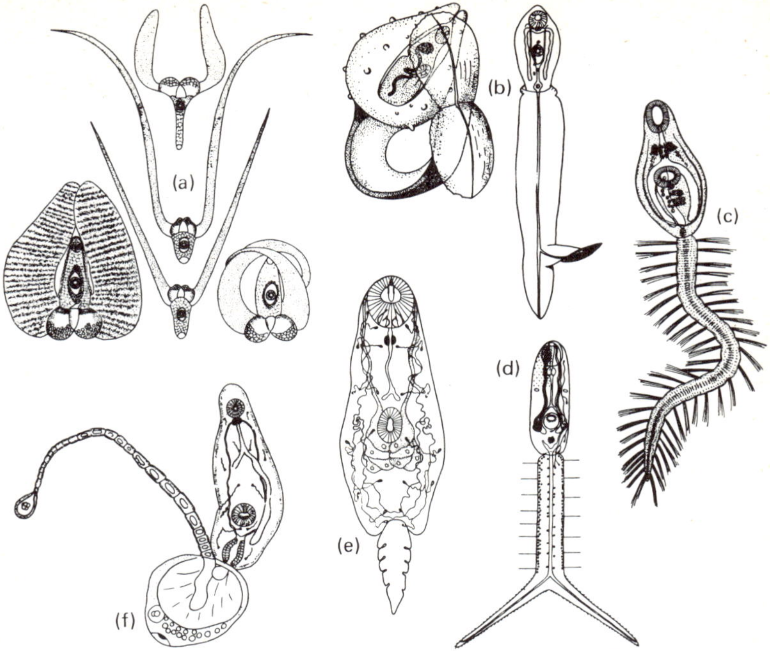

which gives rise to squat cercariae lacking penetration glands but equipped with many cystogenous glands. These are used when the cercaria encysts on vegetation growing near water. When the metacercariae are ingested by the definitive host they excyst and bore through the wall of the intestine, migrate across the peritoneal cavity, and penetrate the liver across its surface. The same kind of life history occurs in *Fasciolopsis buski*, a large worm found in the intestine of man and pigs in Asia; the metacercariae encyst on water chestnuts, amongst other plants, these then being eaten.

(c) Life cycles with two intermediate hosts and with direct development of the cercariae without metamorphosis:

Cercariae need not transform directly into metacercariae but can undergo a different kind of metamorphosis; this occurs in the strigeids.

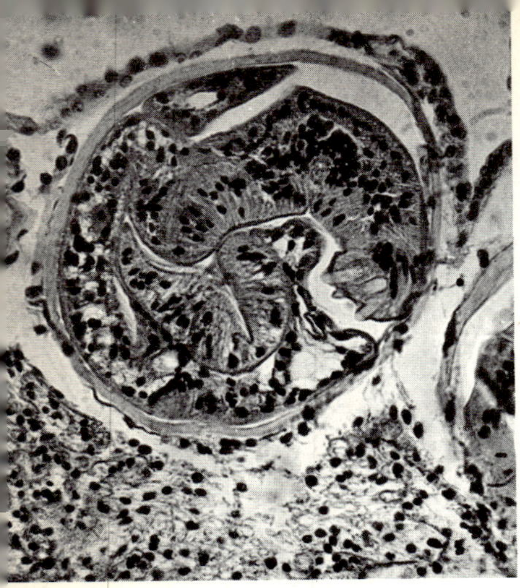

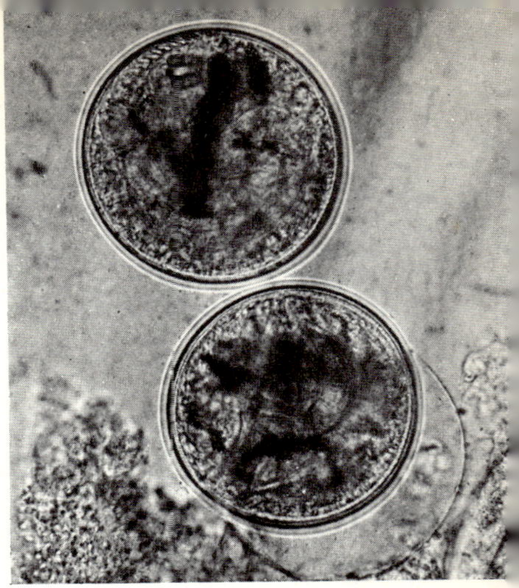

It is better to separate cycles in which cercariae are produced by rediae and those where cercariae are produced directly by sporocysts since this criterion may be of taxonomic significance, even though, from a developmental point of view, the two larval forms are equivalent.

(*i*) *Cercariae produced by a redia and eaten by the second host:* The so-called *cystophore* cercariae (figure 5·33f) are very specialised and have a tail bearing several appendages, with one swollen so as to be able to accommodate the body of the cercaria, which can be withdrawn completely into it. These cercariae swim around freely until eaten by a copepod. From the intestine of this host the cercaria is 'injected' into the haemocoel through one of these appendages. The metacercaria does not encyst. When the copepod is in turn eaten by the definitive host, a frog, metacercariae lodge in the mouth, beneath the tongue or in the eustachian tubes, and here grow into adult worms. The genus *Halipegus*, of which several species are known, has this kind of life cycle, and so probably have the hemiurid intestinal parasites of marine fish.

Lissorchis mutabile is an intestinal parasite of a fresh-water fish and its rediae produce tailless cercariae which are therefore unable to swim. As the cercariae leave the mollusc they are eaten by a small

5·34 Echinostome metacercariae.
Opposite left Section showing the outer envelope formed in reaction to the parasite by the mollusc intermediate host. **Opposite** Metacercaria extruded from this envelope.

annelid, *Chaetogaster*, which lives in association with different parts of the body and mantle of the mollusc. The metacercariae can be seen within the body cavity of *Chaetogaster* and this is eaten, at the same time as the mollusc, by a fish predator.

(*ii*) *Cercariae produced by a redia and penetrating into the second host:* Many examples of this kind of cycle are known and they are all aquatic. It therefore follows that if the second host is a vertebrate, it is likely to be a fish or an amphibian, or if an invertebrate, an arthropod, a mollusc or an annelid. The echinostomes have many species parasitic in birds and in mammals that are tied, either directly or indirectly, to an aquatic environment. That there are a large range of species of echinostomes involved is understandable in view of the fact that the cercariae can penetrate as easily into fish or amphibians as into molluscs.

In *Paragonimus*, a lung parasite of carnivores and man, the cercariae have a penetrating stylet and a short tail. They penetrate the bodies of fresh-water crustaceans, crabs or shrimps. When swallowed by the definitive host, the young flukes traverse the gut wall and migrate directly to the lungs through the diaphragm.

Several trematode families parasitise piscivorous birds and mammals. In the case of opisthorchids which occur in the bile or pancreatic ducts, the cercariae encyst in the muscles of fresh-water fish, especially cyprinids.

Lepocreadium album is a parasite of marine fishes, especially blennies. The cercaria has eyes and is *setigerous*, the tail bearing long rigid bristles. This encysts within nudibranch molluscs which are then consumed by the definitive host.

(*iii*) *Cercariae produced by a sporocyst where the sporocyst is eaten by the second intermediate host:* The life history of *Ptychogonimus megastoma*, a species living in the stomach of sharks, is interesting for several reasons; first because the first intermediate host in which the sporocysts are formed is a dentalid mollusc; secondly because the tailless cercariae do not leave the sporocyst; finally because the sporocysts leave the molluscs to creep around on the bottom of the sea. The second intermediate crab host eats the sporocysts and the cercariae traverse the gut wall and encyst in the body cavity.

(*iv*) *Cercariae produced by a sporocyst where the cercaria is eaten by the second intermediate host:* The eggs of *Gorgoderina*, a fluke living in the urinary bladder of frogs and toads, hatch in water. The miracidium is sucked into small fresh-water pisidid lamellibranchs via the inhalent siphons. The cercariae are equipped with a stylet and penetration glands and have a large tail into which the body can be retracted (figure 5·33e). Depending upon the species in question, the cercariae are eaten, either by a *Sialis* larva or by a tadpole in which they encyst and form metacercariae.

This kind of life cycle also occurs in *Dicrocoelium*, the small liver fluke of ruminants, which is completely adapted to a terrestrial existence. The eggs are eaten by the terrestrial molluscs *Helicella* and *Zebrina* and the miracidium hatches in the stomach, from there migrating to the digestive gland. There are two sporocyst generations and the cercariae, equipped with a stylet and penetration glands, have a long tail. They do not leave the mollusc, however, but enter the pulmonary cavity where they encyst without losing their tails. These cysts irritate the mollusc, causing production of abundant mucus which engulfs the cysts; these are then expelled in packets as the intermediate host glides over vegetation when the ambient humidity is favourable. The mucus allows the mass of larvae to adhere to the grass and protects them to some degree against desiccation. The second intermediate host is an ant which eats the cysts and these hatch within the gut. Only now do the

5·35 Left Miracidium of *Parorchis* already enclosing a redia.

5·36 Right Section through the mesenteric blood vessels of a hamster with three paired schistosomes.

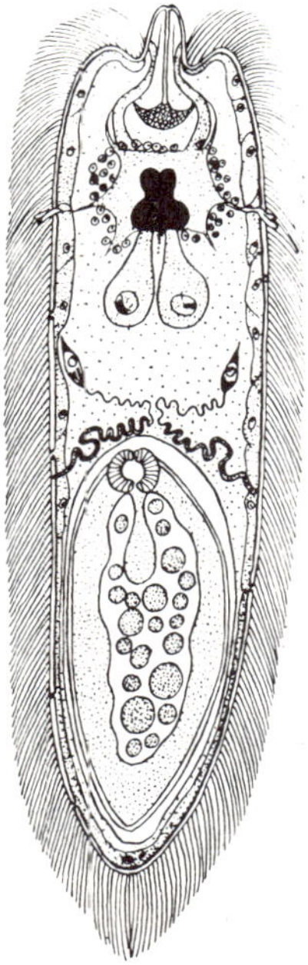

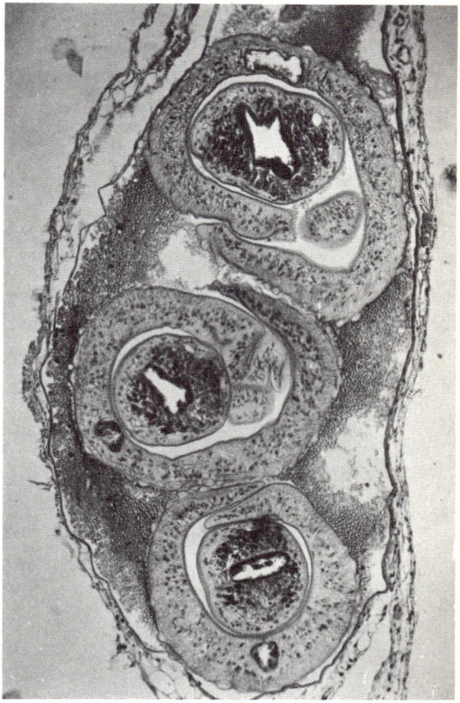

excysted cercariae lose their tails and migrate across the wall of the intestine to encyst as metacercariae in the body cavity. The first cercaria which penetrates into the haemocoele of the ant encysts in the suboesophageal ganglion and this brings about paralysis of the mandibles so that the ant becomes fixed to a blade of grass where the chances of it being eaten by a sheep are greater than if it were to fall to the soil.

In *Dicrocoelioides petiolatum*, a parasite of the bile ducts of magpies, the second generation sporocysts containing cercariae in the snail pass through the lung before leaving the body. The sporocysts are then eaten by wood lice.

(*v*) *Cercariae produced by a sporocyst and actively penetrating into the second intermediate host:* Two situations can be distinguished depending on whether the first host is a lamellibranch or a gastropod.

Bucephalids, which belong to the first category, are parasites of both marine and fresh-water fish, the life cycles being identical for parasites in both these hosts. In *Bucephalus* parasitising carnivorous fresh-water fish such as the pike, the miracidium develops into a ramifying sporocyst in anodontids. The fork-tailed cercariae emitted from the mollusc have enormous tail furcae enabling them to float and to propel themselves in the water. When the cercariae sink to the bottom the tail flukes can become highly contracted to envelop and protect the larva (figure 5·33a). They eventually penetrate into various fish where they become encysted.

When, as in the second case, the first host is a gastropod, the pattern of the life cycle is one common to many trematodes, especially the plagiorchids, which occur as adults in many aquatic or marsh-living birds and mammals. The cercariae are xiphidiocercariae and have a stylet and penetration glands. They encyst in dragonfly larvae or in aquatic dipteran larvae and are thus transported into an aerial environment when these insects metamorphose. Bats acquire the infestation as they swoop low over the water catching insects, and at the same time transmit the trematode eggs back into the water via the faeces.

The cercariae of *Plagiorchis muris*, a common parasite of rodents but also described from a bat, usually encyst in aquatic dipteran larvae but are also able to encyst within the sporocyst without leaving the mollusc, thus encouraging the acquisition of several different kinds of definitive host.

Haplometra is a lung parasite of the frog and its cercarial stage encysts in the buccal cavity of frogs and tadpoles. Excystation soon occurs and the larvae attain the lung directly.

An interesting adaptation of the life history to a terrestrial existence occurs in the brachylaemids. These digeneans occur in birds and small mammals. The first intermediate host is a terrestrial mollusc which is infested passively by eating mature eggs. The miracidium develops into a branching primary sporocyst, which gives rise to secondary sporocysts, which also ramify and which produce cercariae. These have a reduced tail and once outside the mollusc clump together with other cercariae before migrating through the nephrostome into the kidney and pericardium either of the same or another mollusc. The metacercariae do not encyst within the pericardium but remain active. Thus the free-living stage is extremely reduced.

(d) Life cycles with two intermediate hosts where the cercariae develop with metamorphosis:

Only the strigeids have this kind of development. The cercariae, which are produced by daughter sporocysts, are furcocercariae with an anterior crown of small spines which act, together with the penetration glands, to form a highly efficient penetration apparatus. The second intermediate hosts are usually vertebrates – fishes, amphibians or reptiles, and more rarely invertebrates such as molluscs and leeches. On contact with the intermediate host, the cercaria loses its tail and rapidly bores into the tissues where it metamorphoses into a stage equivalent to the metacercariae, a *tetracotyle* (strigeids) or a *diplostomulum* larva (diplostomatids) (figures 5·38f and 5·39). In vertebrates these metacercariae are usually site-specific, occuring for instance in the lens of fish eyes (*Diplostomum spathaceum*), in the rachidian canal of frogs (*Tylodelphys excavata*), in the kidneys of frogs (*Codonocephalus urnigerus*) and

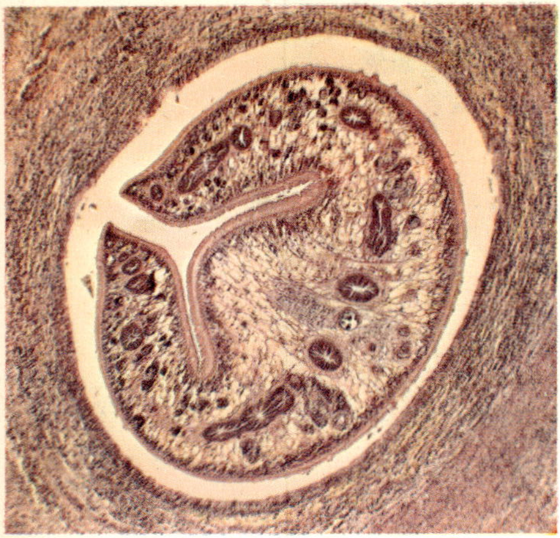

5·37 Left Section through a human liver containing *Fasciola hepatica*.

5·38 Below Metamorphosis of a strigeid. (**a**) cercaria; (**b**) cercaria without tail; (**c**) mesocercaria; (**d**) diplostomulum larva; (**e**) tetracotyle; (**f**) metacercaria.

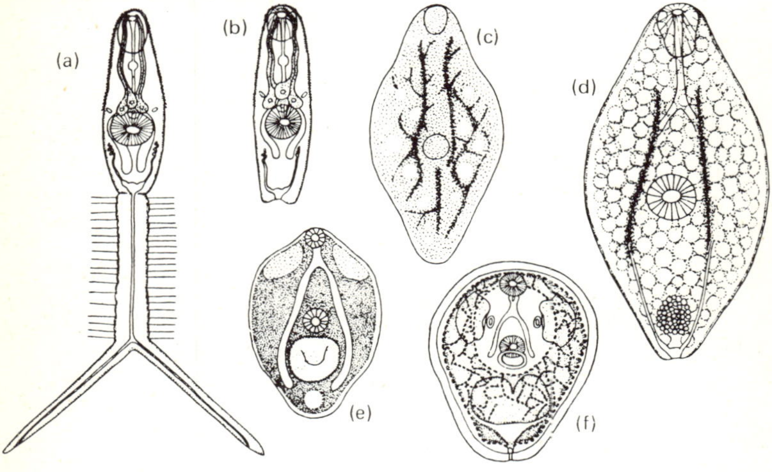

in the skin of fishes (*Posthodiplostomum cuticola*). In this last case melanocytes accumulate around the cyst which is seen by the naked eye to be ringed with black (figure 5·40).

(e) Life cycles with three intermediate hosts:

Again this kind of life cycle occurs only in the strigeids and has resulted by modification of an aquatic life cycle to terrestrial

5·39 Right Metacercariae of *Diplostomum phoxini* in the meninges of a minnow.

5·40 Below Metacercariae of *Posthodiplostomum cuticola* in the skin of bream.

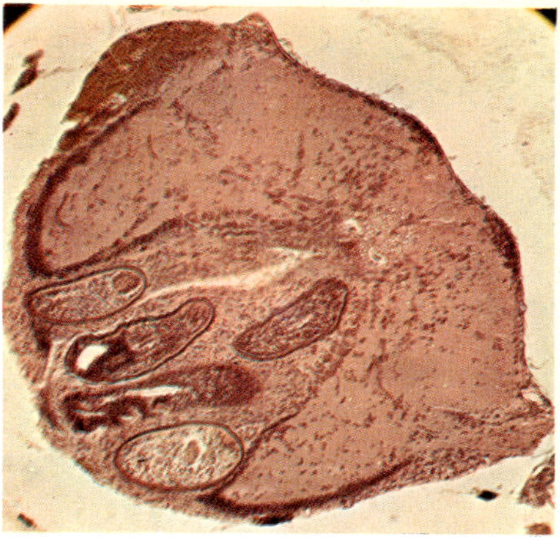

hosts. Two forms of this cycle occur, one in which the third intermediate host is obligatory (strigeids) and one in which it is facultative (diplostomatids).

In *Strigea* the furcocercariae develop in either larval or adult amphibians into *mesocercariae*, an intermediate, non-encysted larval stage which retains the penetration glands. The third intermediate host can be an amphibian, snake, bird or mammal, in which the mesocercaria encysts and transforms into an infestive *tetracotyle*. The definitive host may be either a diurnal or a nocturnal predator or a pillager such as corvids (jays and magpies).

In *Alaria* and *Pharyngostomum* (diplostomatids) living as adults in the intestine of carnivores the cercariae, resembling those described above, penetrate either a tadpole or an adult frog as second intermediate host. The mesocercariae occur in the muscles and are sometimes free sometimes encysted. When the second intermediate host is eaten by a mammal, for instance a rodent, the mesocercariae leave the gut, enter the peritoneal cavity and from here migrate through the diaphragm into the lungs, where the infestive *diplostomulum* stage develops. A carnivore preying on the infested mammal acquires the worm which becomes adult in the gut of the definitive host. This third intermediate host is facultative, since should the final host eat a frog carrying the mesocercariae, these can then penetrate into the peritoneum and pass to the lungs, where an infestive diplostomulum is formed which can escape via the bronchi and trachea back into the intestine to become an adult worm. A similar phenomenon occurs in nematode life cycles such as that of *Ascaris lumbricoides* and is even more striking here since only a single intermediate host occurred in the primitive life cycle that the present one was developed from.

This comparative account of the life cycles of digeneans shows quite clearly that aquatic life cycles are more common in this group than terrestrial life histories which have arisen secondarily from them on several separate occasions. The dicrocoelid parasites of the bile and pancreatic ducts of birds and mammals, for instance, have a terrestrial life cycle, while the opisthorchids which parasitise the same groups of definitive hosts have an aquatic life cycle and so are limited to pisciverous hosts whereas the dicrocoelids are limited to herbivorous and insectivorous hosts. The second intermediate host is of considerable importance in spanning the aquatic and terrestrial environments. Metacercariae encysting in aquatic insect larvae or in frogs or reptiles stand a good chance of infesting a terrestrial definitive host which eats insects, frogs or snakes. The factors involved in distribution of trematodes in their various hosts will be discussed later (p. 177).

Condensation of the life cycle has occurred in various ways and in general is not due to elimination of a larval stage so much as to accelerated development. This is particularly marked when the definitive host is made facultative due to progenesis of the metacercaria rather than when only the second intermediate host is eliminated due to encystation of the cercaria within the sporocyst.

One of the most severe obstacles to successful completion of the digenean life cycle is that imposed by the mollusc on the infestive miracidium. This can be considered as a physiological filter which either prevents or favours development of the parasite larvae. This is bound up with the fact that trematodes are markedly more host-specific to the intermediate mollusc hosts than to the final vertebrate hosts.

6 Acanthocephalans or thorny-headed worms

Unlike the nematodes and platyhelminths, the acanthocephalans, or thorny-headed worms as they are commonly called, are parasitic worms of obscure phylogenetic affinities and cannot easily be related to any other group of invertebrates. They have been grouped, from time to time, with the nematodes, platyhelminths, especially the cestodes, priapulids and even rotifers, but while showing superficial similarities to these groups the acanthocephalans occupy an isolated position which does not allow them to be included with any of them. Their embryology is particularly unusual, commencing with spiral cleavage and followed by breakdown of the cell membranes and syncytial development of embryo. Adult acanthocephalans are likewise characterised by syncytial organisation of the body wall and of most organs except for the gonads and genital ducts.

The acanthocephalans are parasitic throughout life. Their body consists of two regions: an evaginable attachment organ, the proboscis which is always armed with spines, and a posterior trunk region which is flattened and wrinkled in life but becomes more or less cylindrical in water or in fixatives. The body wall is a syncytial structure with few nuclei and is permeated by a system of canals. Beneath this are two layers of syncytial muscles which bound the pseudocoelomic cavity. There is no digestive system. The sexes are always separate and there is a pronounced sexual dimorphism, the females always being larger than the males. The eggs contain an embryo at the time they are laid and the life cycle always involves an intermediate host. Paratenic hosts also occur. Adult acanthocephalans are intestinal parasites of vertebrates. They usually measure about 10 to 20 mm in length, but very large species of acanthocephalans occur in mammals and reach a length of almost one metre. The proboscis is armed with immobile spines which are buried in its wall and this can be invaginated tip first, like the finger of a glove, due to a special system of retractor muscles; when retracted it is accommodated in a pocket with muscular walls

6·1 Male and female *Acanthocephalus ranae*. The copulatory bursa is evaginated in one of the males.

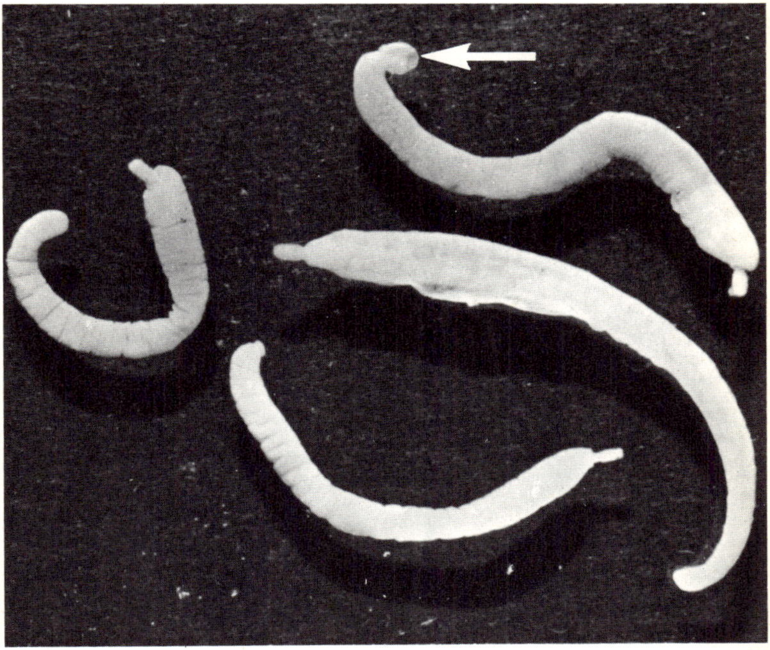

which is quite separate from the pseudocoelomic cavity. As the proboscis spines are backwardly directed, invagination of the proboscis in this manner allows the proboscis to be detached from the intestine without tearing host tissues. Protein- and lipid-rich fluid circulates in the canals, permeating the tegument, and is probably derived from nutritive host intestinal fluids. Despite some histochemical work, very little is known about the nutrition of acanthocephalans. In young forms the gonads are isolated from the pseudocoelomic cavity in two ligament sacs which are attached anteriorly to the proboscis sac and posteriorly at the end of the

6·2 Proboscës of acanthocephalans.
(**a**) *Neoechinorhynchus*;
(**b**) *Illosentis.*

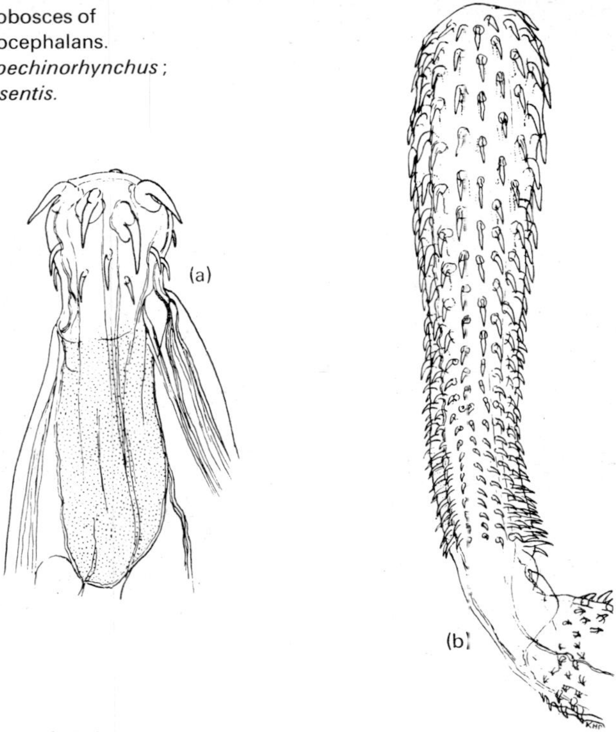

pseudocoelomic cavity. The ligament sacs usually break open when the gonads mature, persisting only as remnants in the form of the ligaments suspending these organs in the pseudocoelom. In one group of acanthocephalans the ligament sacs do not rupture. The female system consists of a rather specialised kind of ovary as well as a complicated sorting organ, the uterine bell, which prevents immature eggs from entering the uterus and being prematurely laid. The male reproductive system consists of two testes, cement glands which help to maintain contact during copulation and also seal the vagina of the female afterwards, and an evaginable bell-shaped copulatory bursa which fits over the posterior end of the female (figure 6·3). Two kinds of eggs are produced according to whether the life cycle is an aquatic or a terrestrial one. In the latter, the eggshell membranes are particularly thick (figure 6·4). The

6·3 Below Posterior region of a male acanthocephalan.
(**a**) Copulatory bursa invaginated;
(**b**) copulatory bursa evaginated
(see also figure 6·1).

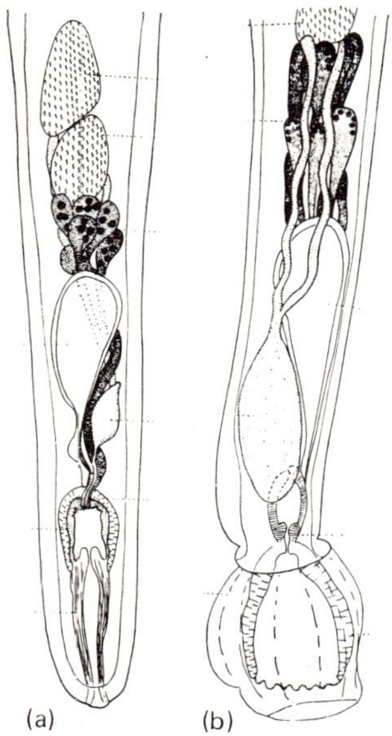

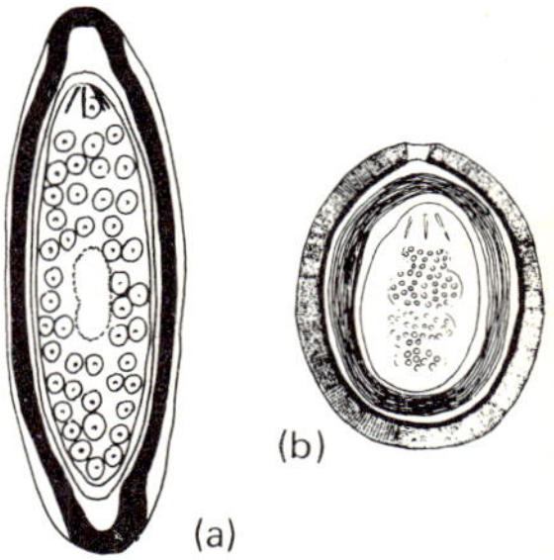

6·4 Above Eggs of acanthocephalans. (**a**) Aquatic type; (**b**) terrestrial type.

mature egg encloses an *acanthor* which has an anterior crown of spines that can be thrust in and out and this assists penetration of the acanthor into the tissues of the intermediate host. The larva also has body spines which facilitate its progress (figure 6·5). This larva hatches within the intermediate host where it develops into an infestive larva. Further development of the acanthor is complicated and is best represented by drawings rather than by description (figure 6·6). The proboscis and its armature develop first, followed by the trunk and reproductive system. At this stage the whole anterior region of the larva is retracted into the pseudocoelomic cavity, also the larva is usually surrounded by a capsule produced by a host reaction to the parasite. The larva is now infestive and can complete its development only in the intestine of the definitive host. In cases where the larva has not reached the

6·5 Below left An acanthor larva.

6·6 Below right Development of the ancanthor into an acanthella in the intermediate host. (**d**) is the infestive cystacanth larva with the proboscis invaginated.

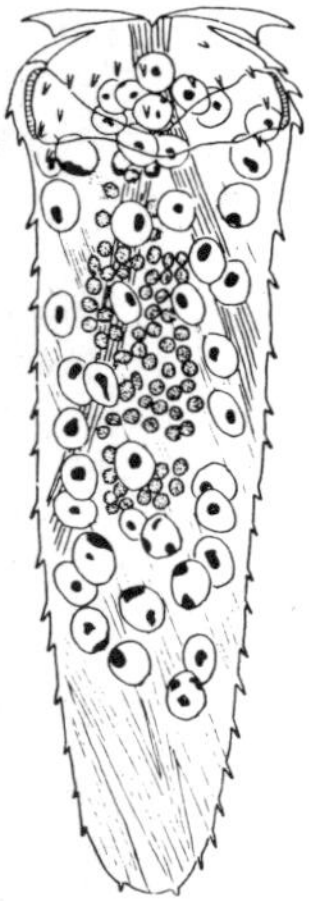

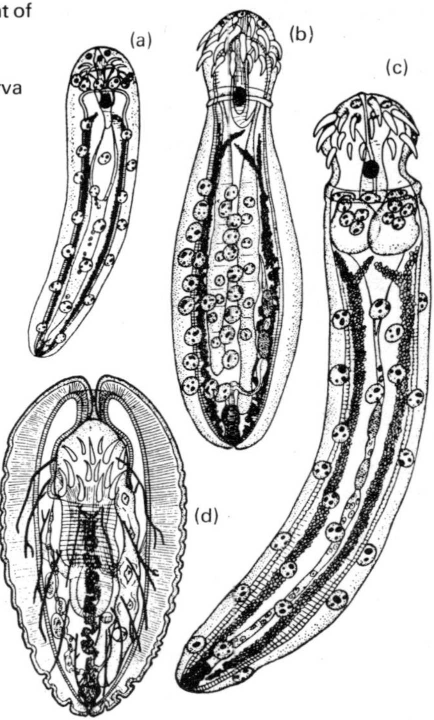

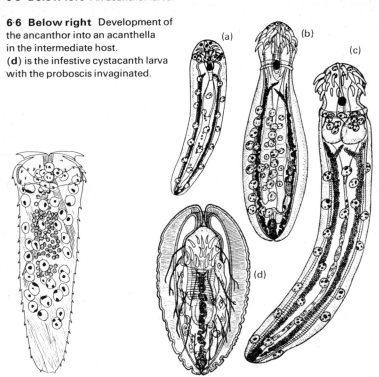

infestive stage before being ingested by the definitive host, this is still capable of migrating out through the intestine of the vertebrate host and completing its development in the peritoneal cavity. Thus this host becomes only a potential intermediate host. This must not be confused with a paratenic host in which it is the already infestive larva that re-encapsulates without undergoing any concomitant morphological or physiological change. It follows from this that a potential intermediate host and a paratenic host may belong to the same species or, even, that the same host can serve both these functions, although this has yet to be established experimentally. Acanthocephalan life cycles are therefore rather more complicated than they might at first appear and there is considerable scope for

6·7 Diagram illustrating the life cycles of acanthocephalans. (A = adult acanthocephalan; I = intermediate host; Inv = invertebrate; juv = infestive larva; ov = egg; Pa = paratenic host; Vert = vertebrate. The paratenic or facultative hosts are shown by a dotted line; the obligatory paratenic host is shown by a solid line.

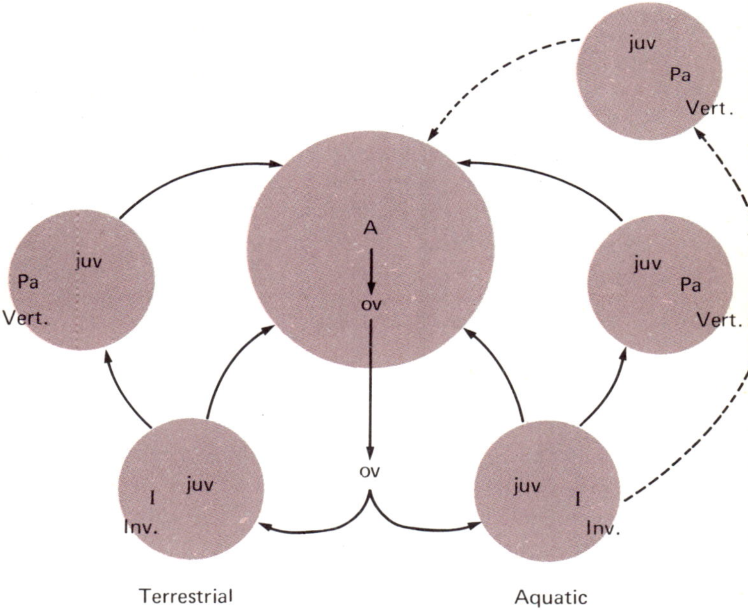

infestation of an extended range of definitive hosts.

For the sake of consistency, it has been decided to arrange the life cycles using the same biological groupings as for the cestodes, though at the same time it is realised that the role of the paratenic host is difficult to assess and that some life cycles are still only incompletely known while many still require elucidation.

Aquatic cycles

These are by far the most common and it seems that acquisition of terrestrial hosts is a fairly recent phenomenon judging from the small number of genera and species that have adapted to them.

(a) With a single intermediate host:

This kind of cycle occurs in acanthocephalans parasitic both in fresh-water and marine fishes. Species of the genus *Echinorhynchus* occurring both in fresh-water and in marine hosts share the same kind of life cycle in that the intermediate host is an amphipod belonging to the genera *Amphithoe* and *Calliopius* for the parasites of gadids, and to *Gammarus* for the fresh-water parasites. Members of the genus *Acanthocephalus* which parasitise fresh-water fish develop in an isopod, *Asellus aquaticus*. *Polymorphus minutus*, a parasite of water fowl, again uses a gammarid as its intermediate host.

(b) With an intermediate host; paratenic host facultative:

Pomphorhynchus laevis occurs in several carnivorous fishes – pike, perch, etc. – and the egg develops to the infestive stage in the gammarid *G. pulex*. The presence of infestive larvae in the body cavity of cyprinid fishes suggests that these are the paratenic hosts which account for the infestation in pike and perch that do not themselves feed on planktonic invertebrates except when young. It is possible that some of these paratenic hosts may really be second intermediate hosts. This has, in fact, been shown to be the case in *Leptorhynchoides thecatus*, parasite of North American fresh-water fishes, in which the infestive larvae are formed after thirty-two days in the amphipod *Hyalella knickerbockeri*. These larva become adult when eaten by young black bass, *Micropterus dolomieni*. If the larvae developing within the intermediate host are less than thirty-two days old when eaten by the fish they leave the intestine to mature in the body cavity. A fish of small size can therefore serve as a second intermediate host for a congeneric host larger than itself.

In *Neoechinorhynchus rutili*, a common parasite of various cyprinids but also pike, eel, salmon and burbot, the egg develops in ostracods which harbour the infestive larva. Among the paratenic hosts involved in this life cycle are *Sialis* larvae (neuropteran), a rhynchobdellid (leech) and a limneid (mollusc), which opens up the possibility of infesting many different kinds of intermediate host.

Members of the genus *Corynosoma* occur in dolphins, seals and marine fish-eating birds. The intermediate host is an amphipod but infestive larvae have also been recorded from many different species of marine fish. In this case the paratenic host has become almost indispensable since there is far less likelihood that seals, cormorants and smews will eat amphipods than that they will feed on fish, which constitute their normal diet.

Terrestrial cycles

(a) With a single intermediate host:

Species of *Moniliformis* occur frequently in rodents, especially rats and the eggs are eaten by cockroaches (orthopterans) and perhaps also beetles, where the infestive larvae develop. There is, therefore, a simple life cycle involving a single intermediate host. This also seems to be true for *Prosthorhynchus formosus* which occurs in a North American woodpecker and undergoes its larval development in a wood louse, *Armadillidium vulgare* (isopod). It is of interest to note that the isopods have both aquatic (asellids) and terrestrial members (wood lice) and that a closely related acanthocephalan *Plagiorhynchus* occurs only in marsh or water birds feeding on small crustaceans.

(b) With an intermediate host; paratenic host indispensable:

In this category are placed the parasites of predatory birds and carnivorous mammals.

Centrorhynchus aluconis occurs in both nocturnal and diurnal predatory birds and infestive larvae have been recovered from amphibians, snakes and insectivorous mammals (shrews and moles). As the infestive larva of a related species has been recorded in an insect (grasshopper) it seems likely that the hosts mentioned above are all paratenic hosts. Predatory birds only rarely eat insects, although some diurnal species occasionally eat grasshoppers, but they do feed on insectivorous mammals, snakes and amphibians. It seems likely then that the first intermediate host is the arthropod which takes in the eggs of the acanthocephalan, while the other hosts are paratenic and not only aid the dissemination of adults in a number of definitive hosts but have actually become indispensable to the completion of the life cycle and therefore to the survival of the parasite.

Although the life cycles of *Oncicola*, *Echinopardalis* and *Pachysentis*, all parasites of carnivores, are not known, they can probably be included in the present category since all the life cycles of acanthocephalans known so far have a first intermediate host which is always an arthropod.

Although our understanding of the cycles and development of the larvae is still very incomplete, and what is known is based upon study of relatively few species, it is possible to approach a review of the cycles as follows. The developmental cycle of acanthocephalans parasitic in fish is simple, with one intermediate host which is generally a crustacean. A second intermediate host could be included in this kind of cycle by accident and would not be necessary for completion of the cycle. On the other hand, in other aquatic life cycles the presence of paratenic hosts has obviously favoured the spread of parasitism amongst carnivorous fish and, following this, in fish-eating birds and mammals. Only in the latter hosts have the cycles become specialised and the paratenic hosts indispensable, coming to play, from a biological though not developmental point of view, the role of second intermediate host. The transition from aquatic to terrestrial conditions already occurred in life cycles using an isopod intermediate host, but could also have happened where infestive forms were harboured by the aquatic larva of a winged insect or in amphibians. It seems true to say, however, that the invasion of terrestrial vertebrates by acanthocephalans is far less spectacular than their conquest of aquatic vertebrates. It could not have occurred at all were it not for the fact that paratenic hosts have been incorporated into the life cycle and have become indispensable.

7 How parasites infest their hosts

The examples of the life histories given in the preceding chapters have emphasised the ecological background against which these cycles are completed, using a succession of different hosts. As the development of the parasite is closely related to conditions within the host it is not surprising that the host-parasite relationship has tended to become highly specialised and that it has stabilised itself at certain levels of infestation. If one were to investigate the parasite fauna, preferably of a defined biotope, over a period of several months, it would become clear that the incidence and degree of infestation in particular individuals was relatively constant over this time and that some animals were susceptible to certain parasites while others were not.

For mallophagans and lice, which undergo their entire life cycle on the host to which the eggs become attached, the problem of locating a new host is not particularly acute for as long as the host survives. In this case intraspecific transfer depends on young animals in the nest acquiring parasites from infested adults; occasionally accidental or passive transfer, for instance by means of an agent such as a winged insect, may occur but the parasites may not survive on hosts acquired in this way. This life cycle is relatively safe, therefore, although adult lice established in host hair or plumage are themselves constantly threatened by the host's preening and scratching. An increase in the numbers of parasites produced helps to compensate for losses due to such hazards and five to fifty eggs, depending on the species, may be laid each day throughout a period of about fifteen days.

Adult fleas are attracted chemotactically to their hosts or to their nests where parasite larvae develop. A single female flea lays only about three hundred to five hundred eggs in all, which is a measure of the comparatively high chances of survival. One safety measure is the ability of a female to survive, but not lay eggs on, an accidentally acquired host.

Pupiparans have a comparatively low fecundity and a larva is produced on average every six or seven days; as the pupal stage intervenes early in the life cycle, however, the dangers of a free-living larval life are minimised. Egg laying occurs in the host nest or on the ground but as the adult fly is not host-specific the first encountered site usually proves suitable. The nycteribids and streblids show a greater degree of site selection and eggs are laid in cracks in caves where the bat hosts live, so that the adult flies, which often lack wings, find little difficulty in establishing contact with their hosts.

Not all life cycles are as economical, and in mollusc, crustacean and helminth parasites a high proportion of larvae may perish, this only occasionally being due to the saturation of hosts causing activation of immune reactions. As has already been established, the reproductive potential of parasites is usually high and eggs are produced often constantly and in large number; a female human *Ascaris*, for example, lays some two hundred thousand eggs per day and has been estimated to produce about twenty million eggs during its total life span. *Diphyllobothrium latum* has been shown to live up to twenty years in man, and as about one million eggs are produced daily this represents a total productivity of at least seven thousand million eggs. Parasitic copepods, isopods and rhizocephalans also produce eggs continuously, unlike the free-living forms of these groups. This contrasts with the insect ecto-parasites, where relatively few eggs are produced but where the life cycle is subject to few dangers.

As has been mentioned, there also exist ways in which the number of larval forms produced by a single egg can be considerably augmented, for instance by budding of the scolex (cestodes) or by polyembryony (trematodes). In certain parasitic nematodes and molluscs accelerated development within the egg or prolonged retention of the egg within the body of the parasite results in the hatching of an advanced larva which stands more chance of survival than it would have done had development occurred in ambient conditions. Pupiparity of the hippoboscids is probably not, however, associated with the parasitism of the adults.

From these observations it can be seen that an enormous number of larval intermediate forms must be lost and that successful encounter with a suitable host at each stage, leading eventually to the final host and sexual reproduction, is a hazardous enterprise. Parasitism is maintained due to compensatory biological processes which confer a decided selective advantage. It would be wrong to believe that successful accomplishment of the life cycle depends solely upon chance, since the evolutionary age of parasites and their modes of transmission has tended to eliminate accident in favour of physiological processes which operate in defined ecological conditions.

Detailed analysis of the ecology and statistical occurrence of, for example, the parasites within a pond would show that they were not distributed in a uniform manner among the different potential hosts present. Consider the food chain of the definitive host of *Diphyllobothrium latum*. The coracidium larva is eaten by the first host, a copepod, to which it is fairly specific. Some physiological process is probably in operation here, preventing development of the procercoid in certain copepods and thus limiting the number of species that can function as first intermediate host for this parasite. This explains why not all the copepods would harbour procercoids. With regard to the second intermediate host, a fish, only those which feed on plankton would contain plerocercoids, so one would not expect to find fish such as carp or tench carrying infestations. Choice of the second intermediate host therefore depends upon ecological factors, in particular host diet. The definitive host is normally a pisciverous vertebrate, and as carnivorous fish serve only as paratenic hosts the choice is narrowed down to warm-blooded vertebrates. Birds, such as gulls and herons, cannot be infested with this parasite, even experimentally, while carnivorous mammals such as the fox, wolf, dog, cat or bear, as well as man, are easily infested. A secondary physiological barrier operating at the level of the intestine is therefore encountered in the definitive host which prevents the plerocercoid developing into an adult worm in hosts to which it is not adapted.

Within this same pond might occur cyprinids parasitised by caryophylleid tapeworms (see p. 126) which use mud-living oligochaetes as their only intermediate hosts. These eat the cestode eggs and oncospheres hatch in the gut and then migrate into the body cavity. In view of difficulties encountered in trying to infest *Tubifex* and *Limnodrilus* with eggs, it seems that here again little understood physiological mechanisms may be involved and that principal and secondary intermediate hosts might occur. Carp and tench, which are mud-searching bottom feeders, are the usual hosts for the adult worm; ducks also scour the mud and feed on oligochaetes but their intestines will not support adult cestodes.

The aquatic molluscs can be divided into those that are bottom dwellers, those that live on vegetation, and those inhabiting the edges of the pond, extending to the limit of the water and waterlogged soil. Each form occupies a distinct ecological niche and by dissecting the digestive gland it can be shown that the same species of mollusc can harbour many kinds of cercariae, each species of cercaria being in general confined to one or several related species of molluscs. In trematode life cycles the first intermediate host, which is the most important from the ontogenetic point of view, becomes infested either by eating embryonated eggs or by being actively penetrated by a miracidium. The fact that eggs are consumed, however, does not necessarily imply that these will hatch in the intestine, nor does it follow that miracidia which have reached the digestive gland can necessarily develop into cercariae. Selective mechanisms – presumably physiological – again operate, allowing only certain species to develop and inhibiting the development of others. Specific processes also operate when snails are actively penetrated by miracidia. These larvae are selectively attracted, by chemotaxis, to those molluscs in which they are able to complete the development. Many examples of this kind of specificity are known. For example, of two species of *Bithynia* living in the same ecological niche, only one is attractive to the miracidia of *Opisthorchis felineus*. Miracidia of *Schistosoma mansoni* are attracted by planorbids but not by bulinids, the latter attracting instead the miracidia of *S. haematobium*. It has been shown that a single species of mollusc can exist as several distinct biochemical races (Wright, 1964). The Irish strain of *Limnaea truncatula* is refractory to infestation by miracidia of the liver fluke but the English strain is susceptible. Sometimes these different strains are in fact separated ecologically and geographically from one another and each harbours different larval trematodes. Thus one can see that a selective barrier exists at the initial stages of the trematode life cycle which must necessarily influence the nature of subsequent hosts. Cercariae equipped with a stylet and penetration glands often encyst in adult or larval arthropods occurring in the same ecological niche as the mollusc first host. Ecological specificity is an important factor and can influence the course of the whole life cycle. Like miracidia, the cercariae can also be selectively attracted to certain hosts. The cercariae of *Glypthelmins quieta* are not attracted to tadpoles but only to adult frogs, a fact which can be demonstrated experimentally using thyroid extracts to produce an accelerated metamorphosis of the tadpoles. These cercariae penetrate via the skin of the young frog, which is much more glandular than that of the tadpole. Comparatively little is known about the mode of action of the penetration glands; the ability to penetrate some hosts and not others may be due to the enzymatic properties of these secretions. Immune reactions are evoked during infestation and could also constitute a selective mechanism allowing some cercariae to penetrate and not others. Cercariae of *Cotylurus flabelliformis* usually encyst in limnaeids but are only able to do so in those snails which do not already harbour sporocysts of this species. In addition, snails already parasitised by echinostome rediae and plagiorchid sporocysts are also sensitised against these cercariae. Thus the mollusc host seems to produce a general reaction to all trematode larvae as well as a specific response to a single species.

The longevity of the metacercarial stage is bound up with the nature of the second intermediate host; if this is a vertebrate, such as a fish or an amphibian, the metacercaria will survive into the next season or even longer. If the second intermediate host is an invertebrate, chances of survival are considerably reduced. The host may live for only a season or could be a larval stage which metamorphoses into a winged insect, for instance, and transports the metacercaria into a different environment. Metacercariae which encyst on vegetation will only be available to infest the herbivorous definitive host until winter, for as long as the vegetation survives, because metacercariae falling to the bottom of the pond with plant remains will no longer be accessible to these final hosts. Occasionally a parasitised definitive host may stop in the neighbourhood of the pond, a migrating bird for example, and may shed trematode eggs into the water. These introduced eggs are extremely unlikely to be able to complete their life cycle in the pond even if they find hosts which could separately support the different larval stages. It seems that trematodes are not markedly specific to their final definitive hosts, so a physiological barrier preventing worms becoming established in the intestine is hardly in evidence and instead flukes show ecological specificity to their definitive hosts, their presence in them being governed by the kind of food these hosts eat. Many kinds of metacercariae, when fed to laboratory animals, will produce adult worms. There are even trematode genera which which are known only from adults reared experimentally in white mice, the real definitive host not having been determined.

Anodonta may occur in the mud on the bottom of the pond, the glochidia larvae they produce being set into circulation by the movements of the fish, where they attach to their fins to develop into small mussels. A high proportion of the fish may be infested and there may be a large number of adult *Anodonta* at the bottom of the pond. If few fish are parasitised this is because fish which have previously been infested have become immune to reinfestation, the degree of immunity being proportional to that of initial infestation.

The reason why a pond was chosen to illustrate the ways in which parasites discover their hosts (and are also 'chosen' by their hosts) is that this is a convenient model ecosystem with clearly circumscribed boundaries. In a lake it would be possible to recognise a number of ecological niches, the limits of which would often be difficult to distinguish. In such a situation the parasitologist might be interested to note that fishes usually living in open water come into shallow water to spawn, thus exposing themselves to infestation from free-swimming parasite larvae which they would be unlikely to come into contact with in deep water. Infestation would, in such a case, be seasonal and would be geared to the reproductive rhythm of the fish hosts. The host reproductive cycle is known to influence the parasites of amphibians. The common frog, *Rana temporaria*, returns to water to spawn in early March and the polystomes which live in the bladder (see p. 106) produce eggs at the same time; their oncomiracidia larvae hatch at about the same time as the tadpoles which they then locate and on which they develop. In the south of France *Pelobates* has two breeding seasons a year, one in spring and one in autumn, and its polystome also lays eggs twice a year, the reproductive rhythm of the parasite being superimposed upon that of its host. The life cycle of polystomes in tree frogs is also synchronised with that of the host (Combes, 1968). *Rana esculenta*, the edible frog, does not breed until May so polystome larvae are no longer available by the time the tadpoles of this species appear; parasitised *R. temporaria* have already left the water by this time. Where *R. esculenta* is not infested with polystomes this is therefore probably due to the late spawning season of this host. In tropical regions such as Africa, polystomes occur in clawed toads, *Xenopus laevis*, which are permanently aquatic. Here the parasites lay eggs continuously and the larvae do not attach to tadpoles but penetrate into the cloaca of the young toads directly. Unlike the frog parasites no neotenic stages occur in *Xenopus* polystomes (Thurston, 1964).

Many cestodes and digeneans produce eggs only at specific times. Some carnivorous fish, such as pike and trout, cease feeding from the time they start spawning until this is finished. This affects cestode parasites in particular and causes detachment and elimination of the strobila from a region behind the scolex; the scolex may also be lost on occasion. Parasite egg production obviously ceases during this time. Salmon migrating from the sea into rivers to spawn also fast during this time and lose their parasites for this reason and not, as is often believed, simply because of the transition from marine to fresh-water conditions. When captured in estuaries they still harbour cestodes, while fish which have undergone a long migration without feeding, for example those which have navigated the Rhine from Rotterdam to Basle, are clear of cestodes.

An analysis of ecological conditions in the sea is obviously very difficult and there is all too little information available. It is possible to formulate an ecological system for estuaries and littoral zones, but little is known about conditions in the open sea. The parasite fauna of estuaries must be euryhaline and able to withstand wide fluctuations in salinity throughout the various stages of their life cycles. Littoral regions, especially those subject to tidal influence, provide many favourable niches in which parasite life cycles can unfold, niches such as the small pools and patches of seaweed which appear at low tide. Specialised conditions of this kind have resulted in exploitation of a single mollusc host by many different species of cercariae. Cable (1956) noted that a mollusc found at low tide on a beach in Puerto Rico harboured up to a dozen kinds of cercariae from five different trematode families. One has only to see the numerous mollusc-eating birds which flock to these pools at low tide searching for food to realise that such habitats literally proffer infested intermediate hosts to the birds, only to be renewed at each high tide. Even so, one must expect to find differences in the helminth fauna of different species of bird, this being dependent on the precise nature of the diet of different birds.

The life cycle of *Cryptocotyle lingua* has already been mentioned (p. 146). Here cercariae are produced by littoral winkles and encounter shoals of young shore-feeding herrings, this second intermediate host then being devoured by gulls. Crabs infested with *Sacculina* also live in these pools and liberate parasite larvae, which have time to locate a new host between tides.

The littoral region might therefore be expected to have a much richer parasite fauna than the open sea, where conditions would hardly seem to favour completion of parasite life cycles. Information about this is scanty and it is often necessary to draw conclusions on the basis of circumstantial evidence about the distribution of parasites in their hosts. With regard to digeneans, it seems that host fish living at comparable depths harbour similar species of intestinal flukes and this is probably because the host ecological niche in this case lies within a horizontal zone of certain depth characterised by more or less constant temperature, salinity and density. Here then the biotope is stratified horizontally and final and intermediate hosts must meet within a defined stratum. Unfortunately this kind of model does not help to explain how the first intermediate mollusc hosts of trematodes, other than pelagic gastropods, become infested. Pelagic cercariae have been recorded from the plankton but few have been identified. An experiment in which two kinds of cercariae taken from the plankton were placed in a nine metre-high tube filled with sea water, showed that after five hours one species reached the suface while the other did not rise above three metres. This experiment was not repeated but did serve to show that perhaps ceracariae emitted from bottom-living molluscs in the sea could become vertically distributed into zones corresponding with the biotopes of different groups of fishes and of intermediate hosts. In this kind of situation the death of a large number of cercariae would be compensated for by the enormous numbers of these larvae produced by the molluscs.

Very little is known about the way in which the monogenean parasites of marine fishes manage to infest their hosts. Being ectoparasites it is quite likely that the monogeneans are potentially more sensitive to fluctuations in external conditions than, for instance, an intestinal parasite. An isolated observation by Nigrelli (1935) did in fact show that a change in the density of sea water from 1·028 to 1·037 killed larvae of *Benedenia melleni* prior to hatching. It is therefore quite likely that horizontal zones of density constitute the biotope within which monogenean life cycles operate and it may be that the direct life cycle can be completed within a particular zone, flotation of the eggs occurring due to their specific density and the arrangement of their filaments.

Llewellyn (1962) noted that in Plymouth the gill-living monogenean *Gastrocotyle trachuri* was essentially a parasite of young horse mackerel not more than two years of age. Until this time the young horse mackerel live in shallow inshore waters, but after reproducing they disperse into deep water where conditions for infestation with *Gastrocotyle* are less favourable. Within shallow waters a periodicity in the incidence of infestation occurs and juvenile monogeneans can be found on the gills from October to June, while adults are found only from July to September. This can be explained in terms of the behaviour of the host fishes. In winter the horse mackerel feed off the bottom, in summer they surface and become plankton feeders. As the eggs of the monogenean sink to the bottom of the water, where the larvae hatch, the fish become infested during the autumn and winter from eggs laid the previous summer.

Terrestrial life cycles can be considered to be of two main types, depending upon whether the first part of the life history remains aquatic or whether terrestrial hosts are used throughout.

The parasites of fish-eating mammals, birds and reptiles tend to have life cycles belonging to the former category. Aquatic birds that migrate can play an important role in the dissemination of their parasites and though there will be many occasions when favourable conditions for further development are not encountered, in some cases the eggs will contact suitable intermediate hosts. In addition, life cycles which involve vertebrates, fish, reptiles and amphibians, as intermediate hosts, stand more chance of being completed, even after a year has elapsed since infestation, than if a short-lived invertebrate such as a crustacean or insect larva is used.

Exclusively terrestrial life cycles are found more frequently among nematodes, cestodes and acanthocephalans than in trematodes, where they occur only as exceptional cases of secondary adaptation (*Dicrocoelium*, p. 156; *Brachylaemus*, p. 159; *Leucochloridium*, p. 148).

With regard to terrestrial vertebrates like the birds and mammals, the concept of biotope can be substituted for that of territory, within which several biotopes may occur. The territory of the lion comprises the river or water hole as well as the savanna where the lion hunts. Many other animals crowd the water hole to satisfy their thirst – ruminants, elephants, monkeys – and the eggs of the parasites they harbour are evacuated on to the damp ground, only to be eaten by ruminants such as antelopes, which in this way acquire the cysticercus larvae of the large taeniid tapeworms of carnivores. The intestinal parasites of ruminants living in herds, on the other hand, are acquired either by the cropping of infested grass (nematodes, p. 83) or by the penetration of nematode larvae via the skin (p. 83), or in the case of cestodes by ingestion of, for instance, oribatid mites infested with cysticercoid larvae which occur around plant roots. Insectivores and rodents which carry larval stages of the parasites of small carnivores will obviously occupy the territory of the latter, as will arthropods, earthworms and molluscs, which are intermediate hosts for the intestinal and lung parasites of these carnivores. One can in fact predict which small carnivores will be present in a limited territory by examination of the larval parasite fauna found in wild rodents. The importance of paratenic hosts, whether birds, mammals, reptiles or amphibians, is shown particularly clearly in terrestrial life cycles, for these vertebrates constitute normal links in the food chain of carnivores while arthropods, though carriers of infestive larvae, are unlikely to be eaten. Furthermore, the peculiar opportunities for infestation in young nidicolous birds which are fed for a while by the parent birds must not be neglected. The term *nidicolous* refers to birds which, when newly hatched, lack down feathers and have their eyes and ears closed; they generally occupy deep nests. Examples are tits and blackbirds. These nestlings have to remain in the nest for a long time, until the feathers are formed and the eyes open; they are fed by the parents during this time. The opposite situation is that of *nidifuges* where the young are born with functional eyes and are covered with down. Here the nests are often at ground level and the young birds leave the nest early and fend for themselves, e.g. partridges and other gallinaceous birds. These terms can also be applied to mammals; carnivores are nidicolous because they bring back prey for their young, rodents are nidifuge.

This is well demonstrated in ducks which nest in marshes and often have highly parasitised young. The same phenomenon occurs in blackbirds, where the young are stuffed full of intestinal nematodes which later are rejected spontaneously (self-cure reaction) (Baer, 1961). Problems connected with parasitism in migratory birds are dealt with in chapter 8 as they are related to host specificity; discussion of factors determining hatching of nematode eggs in the gut of ruminants will also be postponed until chapter 8.

This somewhat necessarily schematic account makes all too clear the gaps which have yet to be dealt with. However, it might be worth summarising what has been said in this chapter while remaining, at the same time, conscious of the shortcomings of the account as rendered so far and trying to avoid the pitfalls of overgeneralisation. It is hoped that this account has helped to clear up some of the mystery surrounding life cycles inherited from the days when the life history of the taeniids of cat, dog and man were not known and when the relationship between intermediate and definitive host was considered to be relatively straightforward. For instance, it used to be thought that *Taenia taeniaeformis* in cats resulted simply from the eating of infected mice and that providential nature could at the same time provide the cat with food and maintain the parasites. This same kind of reasoning, typical of the nineteenth century, applied equally to dogs eating rabbits and man eating pork or beef. P. Abildgaard, working in Denmark, had as early as 1771 elucidated the first life cycle of a worm by managing to infest ducks with the plerocercoid larvae of *Ligula* obtained from the body cavity of cyprinids. He even showed experimentally that it was possible to maintain these larvae in the peritoneal cavity of dogs. These results were to remain buried beneath less important work, however, and their significance appreciated only one hundred and fifty years later.

The analysis attempted above tries to show how the increased reproductive potential of parasites can be considered as a specialisation to the hazards of this way of life. Of a million eggs produced by an adult bothriocephalid, it has been estimated that a mere four will develop into adults, this representing a success rate of 1 in 250,000! Yet this is sufficient to maintain parasitism. The greater the number of hosts involved in the life cycle the greater are the dangers involved, yet the enormous numbers of eggs produced obviously help to counter difficulties of this kind. The larvae of endoparasitic molluscs parasitising holothurians, once released into the sea, have little difficulty in locating another holothurian, so this is a relatively safe kind of life cycle. It seems quite likely that the high preference shown by parasitic molluscs for echinoderms is the result of a selection favouring acquisition of a sedentary host which is easily infested by larval stages, especially as these hosts live in regions where water currents, which would tend to displace the infestive stages, are minimal.

It seems likely that, insects aside, parasitism in most cases has aquatic origins, although some groups of nematodes, such as the strongyles, are an exception to this and have obviously arisen from free-living terrestrial forms. Basic dependence on aquatic conditions is clearly shown in the trematodes, where the few terrestrial life cycles that exist represent secondary modifications of aquatic cycles. The same is true of the acanthocephalans and cestodes although the latter have become highly adapted to terrestrial conditions and have a specialised larva. *All* the infestive larvae of cestodes have the adult scolex already formed; the larval envelopes are essentially trophic structures which absorb nutrients from the intermediate host and at the same time protect the larvae against host tissues. The bladder worm (cysticercus) within its fluid-filled cavity is insulated from the constant muscular contractions of the intermediate host while the cysticercoid, living in the body cavity of invertebrates, can move around freely. There are actually larval forms intermediate between cysticercoid and cysticercus, showing that there is less fundamental difference between these two than might have been imagined.

An aquatic environment is less rigorous than a terrestrial one, in terms of the selection pressure it exerts, and therefore favours the production of new genotypes. It is now recognised that among the nematodes parasitic in vertebrates, the trichurids, spirurids and ascarids arose from fresh-water ancestors, so acquisition of these parasites by terrestrial vertebrates occurred only recently.

Ecological conditions in an aquatic environment are probably comparatively constant and therefore fairly easily analysed. It seems likely that aquatic life cycles originally depended largely on the ecology and food chains of the successive hosts found in the same environment.

Gradually ecological isolation of particular host-parasite systems might lead to physiological specialisation and the development of host specificity would bring about isolation of hosts and lead to host specificity which might involve immunological factors. This kind of 'marshalling yard' for the network of possible life cycles would become progressively established and would lead, to some extent, to the isolation of adult parasites on their definitive hosts. The transformation of an aquatic life cycle into a terrestrial one has occurred several times; it has been mentioned that among the nematodes at least three groups of parasites have become terrestrial; other groups have hardly attempted this transition, for example monogeneans, where the terrestrial habit is exceptional (*Polystoma*).

Terrestrial life cycles are usually abbreviated compared with aquatic life cycles and this may be brought about by loss of an intermediate host perhaps involving migration within the definitive host. In some terrestrial acanthocephalan and nematode life cycles, on the other hand, a paratenic host may become obligatory. This seems to be a secondary adaptation of aquatic life cycles to terrestrial conditions, where the danger that the life cycle may be interrupted is particularly great, judging from the wide range of ecological niches occupied by terrestrial vertebrates compared with those inhabited by fish and amphibians. In the next chapter it will be seen how ecological, and consequent reproductive isolation of the definitive host, and evolution of their parasites are inextricably bound up with one another.

8 Host-parasite relationships

In the preceding chapters dealing with the different kinds of parasite life cycles it was noted that parasites were physiologically adapted to their hosts, as experimental demonstration has shown. It has become obvious, in fact, that most adult parasites occur on hosts or groups of hosts that are related to one another, either in a real phylogenetic sense or ecologically. The occurrence of related groups of parasites on related hosts demonstrates the degree of intimacy between the two; this phenomenon is termed *host specificity*.

To investigate this problem as a whole, information from many fields should be used: statistical data relating to the distribution of parasites on their hosts and data from parasite collections as well as from experimental results. As these various kinds of documentation are not of equal value and as all these kinds of data are not available for many groups it is considered preferable to deal separately with each group of parasites and only afterwards to attempt a synthesis.

The situation in the parasitic molluscs is difficult to evaluate since observations made on this group are somewhat scanty. However, we know that they are all parasites of echinoderms and are mainly endoparasites of holothurians.

There is a considerable literature on the parasitic Crustacea but little information regarding their host specificity. The observation, made by Giard at the end of the last century, that each species of epicaridian had its own host has not been confirmed.

As has already been noted (chapter 3), host specificity in the obligatory ectoparasitic insects assumes two superficially different forms. The associations of fleas and pupiparans with their hosts are governed primarily by ecological factors, but a certain degree of physiological specificity has arisen, especially in fleas where the composition of host blood seems to be capable of influencing the production of viable eggs. The rabbit flea, *Spilopsyllus cuniculi*, produces eggs only after imbibing the blood of a pregnant doe, so here host hormones also control the reproduction of the flea (Rothschild & Ford, 1964). Most fleas occur on hosts living in nests or burrows, like birds or rodents, since their larvae develop in the sheltered conditions offered by such places.

Mammals such as the primates, ungulates and carnivores, which do not make such permanent dens and therefore stand less chance of coming into contact with their parasites, harbour fewer genera and species of fleas and these fleas do tend to produce a comparatively large number of eggs; in the jigger fleas (tungids) the females have become sedentary and burrow into host skin.

Adult fleas living on mammals are usually relatively non-host specific and can live on a range of mammals. However, fleas parasitising den-living conies, foxes and badgers, and even animals such as the bear which occupies a den only at certain times of the year, tend to be host specific.

The nycteribid and streblid pupiparans show an ecological specificity to their bat hosts but because the ecology of the hosts is so specialised, different species of bats occupying different situations within the same cave, a fairly exclusive kind of host specificity has developed. The relationship between hippoboscids and their hosts has also been shown to be predominantly ecological (Bequaert, 1953–4; Theodor, 1957). Despite this there is little evidence that exchange of bird and mammal ectoparasites could occur and in general forms parasitic on mammals show a more marked host specificity than bird parasites. Members of the subfamily *Melophaginae* occur only on ruminants, the *Alloboscinae* occur on primates, while the *Ortholfersiinae* are specific to Australian marsupials. In general about sixty per cent of known species are specific to related groups of hosts and the most highly host-specific forms tend to occur on hosts that are themselves highly specialised and ecologically isolated. Climatic factors can influence infestation in certain cases; reindeer inhabiting subarctic and arctic regions, for instance, are not parasitised by hippoboscids and this is probably due to the fact that the herds are continually on the move and do not remain long enough in contact with the developing larvae to become infested with adult flies. It was found, however, that a herd of reindeer introduced into Scotland from Norway became infested with the deer parasite *Lipoptena cervi* after only a few weeks (Bequaert, 1954).

The pupiparans have not become physiologically host specific because, although blood feeders, they harbour yeasts and symbiotic bacteria which make them to some extent independent of deficiencies in various kinds of blood diet; instead they exhibit a marked ecological specificity to their hosts.

The phthirapterans are highly host specific and the specificity of the blood-sucking anopleurans seems to be physiological. Unfortunately the taxonomy of sucking lice is at present in a highly confused state, due to earlier inaccuracies in identification (Hopkins, 1957). The absence of anopleurans from Australian marsupials can be explained by the fact that this continent became separate at the beginning of the Tertiary, before lice had colonised the mammals of neighbouring land masses. The presence of *Microthoracius* on both North African dromedaries and South American llamas provides strong evidence that North American camelids were parasitised by this kind of louse, or by its ancestors, as early as the Eocene. The lice of pinnipedes belonging to the family Echinophthiriidae are interesting in that they are a very ancient group which probably originated in the Oligocene. The absence of phthirapterans from bats can perhaps be explained by the fact that many other haematophagus ectoparasites, such as the fleas, nycteribids and streblids, were already established on these hosts. The anopleurans of primitive mammals, such as insectivores, were probably acquired secondarily from rodent parasites and secondary adaptation to other groups of mammals must also have occurred. Despite the confused state of anopleuran taxonomy it does seem that there is a fair correlation between the distribution of related parasites and the phylogeny of their hosts.

Mallophagans are highly specific to their bird hosts and have been made the object of detailed investigations (Clay, 1957). The food of these chewing lice consists essentially of keratin derived from fragments of skin and feathers, but blood is sometimes ingested. This kind of diet is probably much more specialised than was once believed. Thus while it is possible to culture chicken mallophagans *in vitro* under conditions of appropriate temperature and humidity, using dried chick blood and feathers as a food supply, it is not possible to substitute heron feathers for chick feathers without the mallophagans dying. The structure of the feather may also be important. The extent to which interspecific transfer of mallophagans between their bird hosts is possible has already been reviewed (p. 72). It has also been noted that birds of prey may acquire the feather lice of their victims. Most orders of birds, however, have an endemic mallophagan fauna of their own. Certain species occurring on passerines are shared by martins, trogons, rollers and woodpeckers. Clay (1957) has attempted to classify the bird hosts on the basis of the mallophagans they harbour. The result differs greatly from the classification suggested by Mayr & Amadon (1951) but is very like that established by Chandler (1947) on the base of feather structure. Thus the kind of results cited above suggest that physiological specificity exists, which gives the impression of phylogenetic specificity without it being possible to establish definite parallel evolution between the mallophagans and their bird hosts.

The monogeneans are also ectoparasites; they live on fishes and have a direct life cycle. These kinds of conditions provide good material for studying the relationship between isolation on an individual host, or group of hosts, and speciation.

In theory, the viviparity of gyrodactylids would make them excellent material for this kind of speculation (p. 104) because it results in the production of genetically identical clones on a single host. Recent work has demonstrated that a single species of *Gyrodactylus* is associated with a single fish host and here lives on the skin, fins or gills which constitute ecological microhabitats. A fish may carry several species of *Gyrodactylus*, but these differ from those harboured by even closely related fishes, *when both hosts occur in the same habitat*. Furthermore, it is impossible to break down this host specificity and cross-infest different species of fish with the same species of parasite under experimental conditions (Malmberg, 1964).

All other monogeneans are oviparous and have a free-living larval stage. The cyprinids are one of the most important fresh-water fishes found in the northern hemisphere and Africa. Collectively they support some two hundred species of monogeneans, but each of these is confined to a particular host. Where several species of fish bear the same species of *Dactylogyrus* this is invariably because these hosts are capable of forming hybrids. This illustrates the way in which closely related hosts are able to offer their parasites similar ecological niches and similar physiological conditions. In the case of *Diplozoon paradoxum*, forms specific to the roach cannot be established experimentally on bream but can be established on the roach-bream hybrid, while those usually associated with bream can be transferred successfully both to the roach and to the hybrid (Bovet, 1967).

The way in which parasite larvae are specifically attracted to their hosts has been investigated in a few cases. The oncomiracidia larvae of *Entobdella soleae*, for instance, are attracted to their sole host (*Solea solea*) by a specific substance secreted by the skin of the fish; once having contacted the fish the larvae immediately become attached and shed their ciliated coat. Agar impregnated with this substance also exerts an attractive influence. The chemical nature of this attractant is not known but it certainly appears to be specific to the sole because larvae experience little or no attraction to other species of sole or to plaice under the same experimental conditions (Kearn, 1967). *Benedenia melleni* has been recorded from the body surface of at least fifty-seven fish species belonging to fifteen families. It tends to occur on the cornea, which is not vascularised and therefore not able to exert an immune reaction, so parasites are tolerated by a wide range of hosts. In artificial conditions, for instance in aquaria, it can be seen that some fish are naturally resistant to infestation and that others become immune following the first infestation. Mucus scraped from immune fish and placed in water kills monogeneans introduced into it. The same result occurs with mucus obtained from fish that never harbour *B. melleni*, but these monogeneans are not adversely affected by mucus from a normal, non-immune host.

These few experimental studies support statistical evidence showing that monogeneans are strictly separated on to their different hosts and demonstrate the existence of a biochemical barrier to infestation which must have arisen in the course of host evolution and would therefore be independent of the ecological conditions in which the life cycle occurs.

The life cycle of acanthocephalans is largely determined by ecological factors, but signs of specialisation to particular intermediate hosts can nevertheless be recognised. The acanthocephalan *Polymorphus botulus* lives in the eider duck and its larvae occur in a littoral crab under natural conditions. Attempts to infest gammarids obtained from the same ecosystem are not successful, despite the fact that in the Soviet Union larvae of this parasite do occur in gammarids, so this could be due to local or regional factors (or due to the parasites belonging to different races). Other species of *Polymorphus* normally use gammarids as intermediate hosts. *P. minutus*, for example, will develop under experimental conditions in *Gammarus pulex*, but the eggs are destroyed when eaten by *G. duebeni*, a less ubiquitous species than the preceding one and probably more specialised with regard to its habitat. In the British Isles there are three races of *Polymorphus minutus*, each adapted to a particular species of gammarid (Hynes & Nicolas, 1958). So it may be that this acanthocephalan, which has been recorded as an adult from eighty-four different species of birds and is therefore apparently not specific to them, might exist as several larval races, this controlling the ecological distribution of the adult worms. Similar observations have been made regarding the intermediate hosts of several terrestrial life cycles and some hosts have been shown to be much more favourable than others (some of which may prove positively unfavourable). Thus although some degree of physiological host specificity may apply to the selection of the intermediate host, the adult worms show an ecological specificity to their definitive vertebrate hosts.

In nematodes that have been comparatively little modified by parasitism, such as the intestinal strongyles, where the ensheathed third-stage larvae are passively eaten by the definitive ruminant host, a lytic secretion issuing from the excretory pore and aiding exsheathment has been demonstrated. This exsheathing fluid appears to be produced by the excretory system in response to a host stimulus consisting of an increase in carbon dioxide tension, the larval cuticle being permeable to this gas. A similar response brings about hatching of mature eggs of dog, cat and human ascarids (Rogers & Sommerville, 1963). These isolated observations are comparatively limited and should not be generalised, but they do indicate a possible mechanism within the digestive tract of ruminants which could determine the distribution of different strongyle species within the gut. Similarly, the presence of a hatching factor for *Ascaris* eggs may help to explain the kind of intestinal barrier that prevents hatching within the intestine of hosts to which the parasite is not adapted. Much more could be done to test such hypotheses experimentally, in view of their general significance.

Skin penetrating larvae such as those of *Ancylostoma* (p. 83) also encounter a host barrier, the nature of which is not known, preventing larvae penetrating further than the epidermis in abnormal hosts. Larvae of *A. caninum* from dogs and *A. ceylanicum* from carnivores, for instance, are unable to penetrate beyond the skin of man and here produce a rash known as cutaneous larva migrans or 'creeping eruption'.

In East Africa trichinosis is propagated between large carnivores and bush pigs but has never been recorded in the wart hog, which is the favourite prey of lions and is more widespread than the bush pig. This is probably due to the feeding habits of the wart hog, which eats roots rather than scavenging on the bodies of dead animals. The absence of trichinids from wild rodents, of which over a thousand species have been examined, is probably attributable to the same kind of reason. Laboratory tests show that African strains of *Trichina* become established in rats and mice with much greater difficulty than the European strain; they seem to have undergone a physiological modification which is probably more ancient than the recent discovery of this African strain might lead one to suppose. Sylvatic trichinosis is especially adapted to the mammals and ecological conditions of East Africa; this has given rise to a type of local host specificity which in turn has led to production of specialised races of parasites. The microfilaria larva of *Loa* parasitic in man shows a diurnal periodicity and is transmitted by species of day-biting *Chrysops*. There is also a strain of this filaria worm in monkeys which has a nocturnal periodicity and is transmitted by species of *Chrysops* which are active only in

8·1 Example of 'pyramidal' specificity in the life history of *Orneascaris robertsi*, a parasite of pythons.

the evening and the early part of the night; also they are confined to the tree tops and do not bite man, so the two races of parasite are kept separate because of the separate ecologies of their intermediate hosts. In general, nematode larvae are not particularly specific to their intermediate hosts; *Cheilospirura hamulosa*, a parasite of various birds, passes indifferently through crustaceans (amphipods), orthopterans (grasshoppers) and beetles. Chabaud (1965) has suggested, however, that a certain selection of intermediate hosts has occurred, in that the least specialised nematodes tend to utilise intermediate hosts with a slow metabolism while more specialised forms use intermediate hosts with a more elevated metabolism (such as dipterans). Over and above this basic situation, all variations ranging from low specificity to a very rigid specificity to the intermediate host are observed. Experimental work on the life cycle of *Orneascaris robertsi*, an intestinal parasite of an Australian python, has shown that considerable flexibility is possible in the life cycle; there may be only a single intermediate host, or a series of successive hosts can intervene at the level of the first larval stages (Sprent, 1963). Thus there is a kind of 'pyramidal specificity' which is essentially physiological, the life cycle having a broad base represented by the invertebrate and vertebrate hosts which can eat the eggs and harbour second-stage larvae. In the middle region of the pyramid only mammals occur, these having acquired the infestive larvae by eating the intermediate hosts of the stage below. At the summit of the pyramid occurs only a single host, which is the definitive host (figure 8·1). It is curious that the life cycle which at first seemed so favourable for dispersing the larvae into many different hosts should terminate finally in a single host. This kind of pyramidal specificity illustrates particularly well how important are the ecological conditions in which life cycle operates. Finally it is worth mentioning that laboratory rats and mice which do not form part of the Australian fauna are less

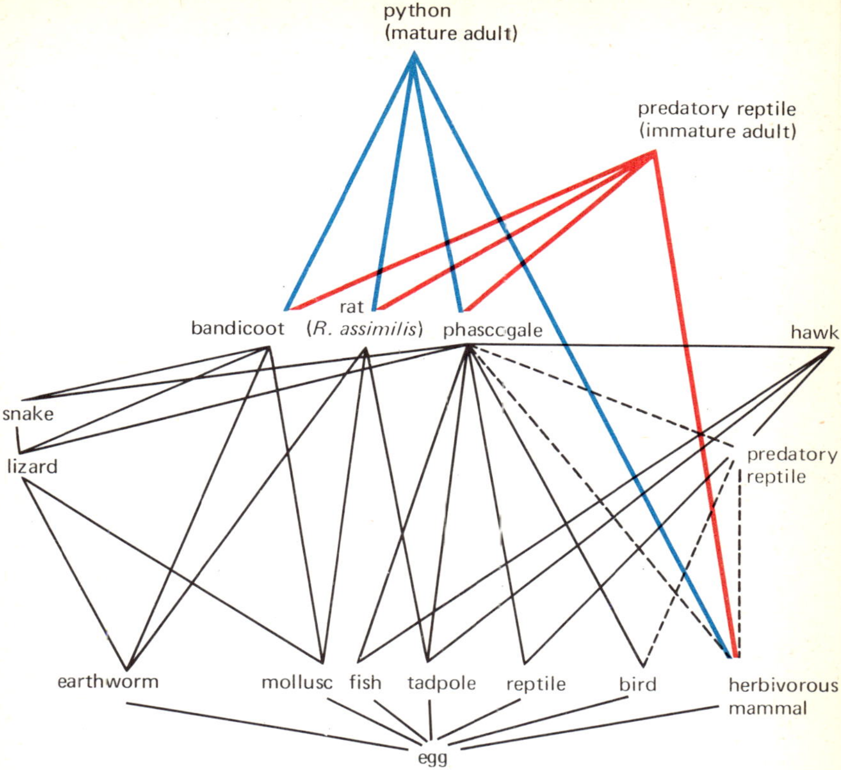

favourable hosts for larval development than the native rodents.

An investigation into the host specificity of adult nematodes is complicated by the polyphyletic origin of nematode parasites and also, on occasion, by the difficulties of precise identification of the worms. In most cases host specificity is both physiological and ecological in nematodes. Some authors (Osche, 1958), however, suggest that the Ascaroidea show phylogenetic specificity, which would mean that they were very early parasites of vertebrates and had evolved in concert with them. Morphologically 'primitive' forms of ascarid occur in the more primitive vertebrates while more highly differentiated parasites are often found to be associated with the higher vertebrates. This basic plan tends to be masked, however, by the secondary acquisition of hosts that cannot be related to a phylogenetic scheme. It is difficult to demonstrate the kind of physiological factors that operate in the intestine of the definitive host, either preventing exsheathment of the infestive larva or inhibiting sexual development, thereby selecting which sorts of adult nematodes may live there. Cases of immunity to new infestations or as a result of superinfestation have often been recorded, but this is essentially a phenomenon concerning the strongyles of domestic animals, where artificial conditions resemble those of a laboratory rather than those occurring in wild animals in nature. Because the life cycles of the nematodes of domestic animals are usually monoxenous, the pasture on which the animals are confined rapidly becomes a kind of culture medium for the larval stages. Under these conditions it is hardly surprising that immunity develops but it seems less likely that this phenomenon would play such a powerful selective role and tend to eliminate some species rather than others, in wild animals under natural conditions. Okapis captured in the forest and apparently in good health have been found to harbour incredible numbers of *Okapistrongylus;* such cases may well be exceptional, however, and could be related to the peculiar ecological conditions of a dense tropical forest.

With regard to trematodes, these show a fairly high degree of specificity to their first, molluscan, intermediate hosts (p. 163), so that these effectively segregate different kinds of cercaria into differing ecological conditions, this in turn influencing their distribution into their second vertebrate or invertebrate hosts. The infestive metacercaria larva is already a small adult in which adult features can be recognised. There is no metamorphosis, the larva merely growing to the adult stage and becoming sexually mature. Occasionally, in strigeids for example, the infestive larvae develop rapidly to the adult state and start to produce eggs after only a few days in experimental hosts, the adult worms being expelled spontaneously shortly afterwards. Several metacercariae of bird trematodes can develop to maturity in white mice, while species normally living in the intestine of gulls can become adult in chickens. There are even forms which are known purely on the basis of adults reared in laboratory animals and for which the real host is unknown.

There is both statistical and experimental evidence showing that adult trematodes are not particularly specific to their final hosts; in fact it is the feeding habits and ecology of the definitive host rather than host phylogeny that determines which trematodes they harbour. The effects of the ecological distribution of fish hosts on their parasites can be quite far-reaching; in tropical seas the trematodes of bottom-living fish are quite different from those of fish living near the surface and are much more closely related to those parasitising cold-water fishes. Thirty-one species of trematodes recorded from fish of northern and southern temperate seas have been found to resemble those occuring in hosts living in deep tropical waters. However, a comparison of the trematode fauna of the Pacific eel and the Atlantic eel shows that while the first harbours eight endemic species the second is parasitised by twenty-one species, only one of which is endemic and differs from those of the Pacific eel. It is interesting to note that no trematodes are known to occur in the spiral intestine of selachians; those living in the stomach or gall bladder are not primitive but belong to groups which are better represented in bony fishes.

Frog trematodes seem to be fairly specific, but statistical data like that relating to bird and mammal parasites apply only to a generalised situation. When the trematode fauna of a defined region is analysed, a certain degree of specificity to the definitive hosts does seem to occur, but this is exaggerated because only a limited sample of hosts are examined and because of a bias imposed by the selective feeding habits of the host which, being confined to a particular region, has a limited choice of diet unlike, for instance, a migratory bird which may feed in diverse areas where the trematodes may also be very different. Fresh-water and marine-life cycles may coexist side by side in the trematodes; migratory fish such as salmon smelt could become parasitised by metacercariae in fresh water and carry them into the sea where sea birds, such as the auk or gannet, could become parasitised by the same species of trematode found in fresh-water birds. Mullet may play a similar role since this is another fish that spends some of its life in fresh water before returning to the sea.

The Strigeida are very specialised trematodes and have a feeding and attachment organ termed a tribocytic organ which is used to lyse the host intestinal mucosa before feeding. It might well be thought that such specialisation in feeding habit might be associated with pronounced host specificity, but it seems that here too host diet is a more important factor in determining its parasite fauna than its relationships within the vertebrate kingdom. *Cotylurus* occurs in ducks, seagulls and grebes as well as many small waders (Charadriiformes). However, a more detailed analysis shows that the species occurring in gulls, grebes and diving birds has an infestive larva in fish while those parasitising ducks and the small waders use a fresh-water mollusc as second intermediate host. The same is true of the genus *Apatemon*, where infestive larvae of species parasitising diving birds occur in fresh-water fishes, while those in ducks and swans occur in leeches. Bird migration complicates the issue further; a marine duck such as the eider often overwinters on European lakes where it can acquire fresh-water duck trematodes. There is some evidence that this group of trematodes can also complete its cycle in the sea but the facts are not yet sufficiently clear to determine whether all species, or a few only of certain genera, are marine throughout their life history. For example,

Cardiocephaloides physalis is a parasite of penguins, cormorants and puffins, which all feed on marine animals, particularly fishes. The other species of this genus parasitise gulls and terns and probably have a marine life cycle. Of the six species of the genus *Mesostephanus*, four are parasites of pelicans, frigates, gannets, cormorants and sea-fish-eating eagles; the other two, however, occur in dogs, in Texas, Rumania and the Ukraine! Which suggests either accidental infestation via marine fishes or that members of a single genus can have both marine and fresh-water life cycles. Some strigeid species from birds can infest dogs in the laboratory.

This problem can be complicated further by the difficulty of identifying the adult trematodes, since they vary in size and in other ways in different hosts; in addition very little is known about the basic physiological requirements of metacercariae needing to be satisfied before an adult laying viable eggs can be produced. Most trematode parasites of man in various latitudes are also shared with rodents, ungulates, suids or carnivores, which can therefore act as *reservoir hosts*. Perhaps only the urinary and mesenteric schistosomes are truly parasites of man, although these parasites can easily be maintained in laboratory animals by allowing cercariae to penetrate through their skin. Although in most cases the evidence points to the fact that the definitive hosts of trematodes are determined by the kind of infested food they eat, rather than these being actively chosen by the parasite, there may be just a few examples where specificity to the final host is determined by a physiological barrier operating in the gut. Unfortunately, experimental work has not furnished sufficiently definite evidence for this kind of barrier; failure to infest a certain host does not constitute proof that this exists.

The relationships of cestodes with their hosts are much better understood, even though the larval stages, the pleroceroid, cysticercoid or cysticercus have to undergo metamorphosis in the intestine of the final host. It seems that the digestive juices bring about excystment, after which the scolex attaches to the gut wall; strobilisation then begins, followed by progressive maturation of the reproductive organs. There is therefore a latent period following excystment when the larva has to absorb nutrients across its body wall to build up reserves necessary to the heightened metabolism accompanying segment formation. This initial growth phase, which varies from a few days to weeks, depending on the species, is therefore a critical period in the life history and during this time the cestode is subjected to the extremely specialised conditions within its host; if the larva is not adapted to these conditions it will be expelled without having reached sexual maturity. Assuming that cestodes are shown to have a strict specificity to their final hosts, one need look no further than this to see how a physiological barrier to establishment of nonadapted larval stages might work, for the larval stages are not as specific to their intermediate hosts.

Every class of vertebrates does in fact harbour a characteristic cestode fauna and each order within this tends to have cestodes special to itself. Sharks, for example, are parasitised not only by different species but also by different genera of cestodes from those found in rays. It could be argued that this difference was produced by the different habits and feeding of these two elasmobranchs, as rays are bottom dwellers. One finds that the different groups of rays – the rays proper, the sting rays and the electric rays – all have their own characteristic species and genera of cestodes, in spite of the similarity in feeding habits. Among the reptiles, the related cestode genera *Duthiersia* and *Bothridium* occur in varanids and pythons respectively and the discontinuous geographical distribution of these forms over three continents suggests that they are extremely ancient parasites which already existed before the hosts had differentiated from other reptiles. The specificity shown by cestodes to bird hosts was first noted about a century ago and has since been confirmed by many workers. Each order of birds has its own cestode fauna and it is striking to note that orders such as the grebes and divers or the diurnal and nocturnal predators, although sharing the same kind of diet, harbour their own characteristic cestodes. Birds of the open sea, such as petrels, albatross, fulmars, penguins, frigate birds, cormorants, pelicans, gulls and auks, which represent four distinct orders, are all parasitised by the genus *Tetrabothrius;* however, each order has its own particular species of this cestode genus. The same kind of separation into different groups of hosts occurs in the mammals, although there are fewer very specialised forms. The genus *Hymenolepis* in insectivores, for instance, has different species in hedgehogs, moles and shrews respectively. Within the soricid family host specificity is developed to the point where white-toothed shrews (*Crocidura*, *Suncus*) and red-toothed shrews (*Sorex*, *Neomys*) harbour different species of *Hymenolepis*, this difference being unrelated to ecological factors. The cestode fauna of ruminants is difficult to analyse, since the introduction of sheep and cows into most habitable regions of the world has resulted in the introduction of new cestode species which can then pass into wild animals and vice versa. The genera *Stilesia* and *Avitellina*, which occur in antelopes in Africa and Asia, have been acquired by sheep and even by the domestic cattle of these countries. These are therefore man-influenced transfers of the kind that have occurred between the dog and cat and other domestic animals. In general, however, each systematic category of hosts not above the level of an order or below the level of family harbours its own particular cestodes. Naturally the local distribution of cestodes depends upon the ecological conditions bringing intermediate and definitive hosts together in the same biotope; however, the degree of infestation of different groups of hosts with different worms depends upon the intimacy of their physiological adaptation to the conditions encountered in the gut of their particular hosts.

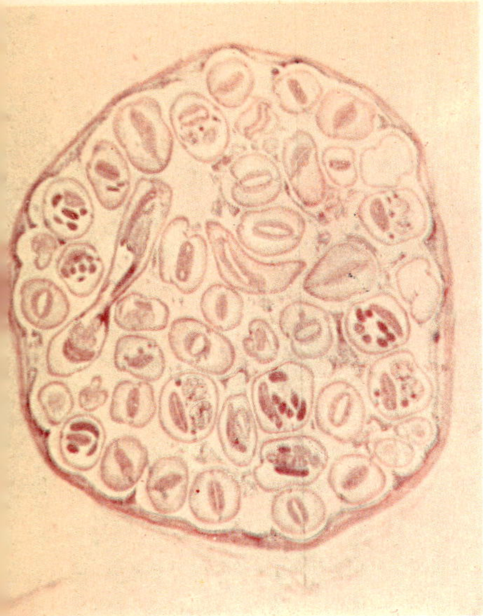

8·2 Left Intestine of an owl, containing nematodes; **opposite** (**a**) scolex with tentacles of the tapeworm *Polypocephalus* inserted into the intestinal mucosa of a ray; (**b**) lung of a sheep heavily parasitised by the nematode *Cysticaulus*; (**c**) scolex of *Anomotoenia constricta* within the intestinal mucosa of a blackbird; (**d**) tribocytic organ of *Pharyngostomum cordatum* which has destroyed the intestinal mucosa of a leopard.

The pseudophyllidean cestodes differ somewhat from the cyclophyllideans with regard to the way in which they are adapted to their hosts, because the plerocercoid larvae of the former can apparently form larval races, in particular fishes; thus there are forms of *Diphyllobothrium* capable either of infesting both fish and mammals, including man, or only either fish or mammals. This same genus also has several marine representatives which parasitise seals and sea lions and can infest man secondarily. This seems to be an instance where there is a basic ecological specificity which has subsequently shown secondary physiological adaptation to intermediate as well as definitive hosts. Young hosts are usually more susceptible to infestation than older hosts; their physiological barrier may be less impenetrable to infestation. In young nestlings a high degree of infestation may be acquired from the diverse food material carried into the nest by parent birds, but this high infestation rate does not necessarily persist into the adult state. For example, *Diphyllobothrium oblongatum* parasitises the chicks of gulls and terns but the infestation is absent from adult birds. This

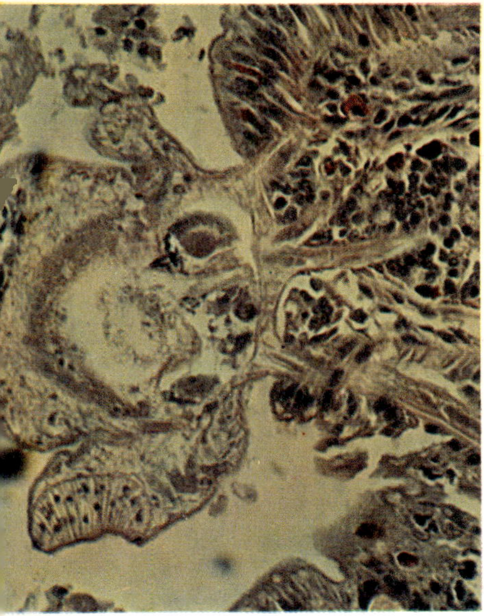

(a)

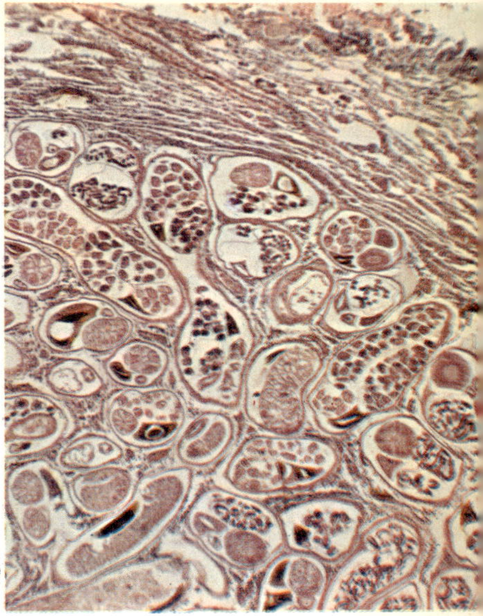

(b)

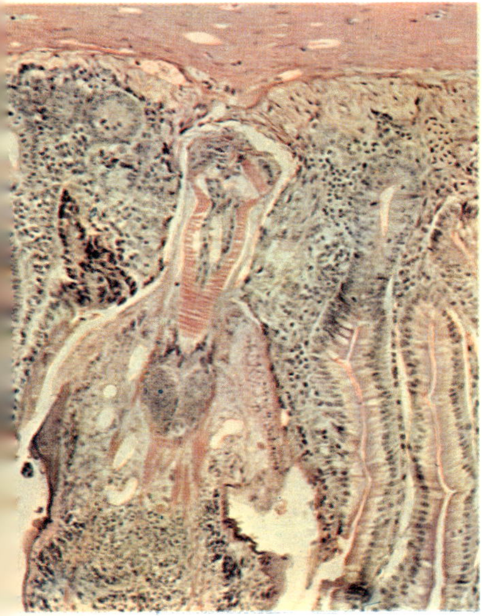

(c)

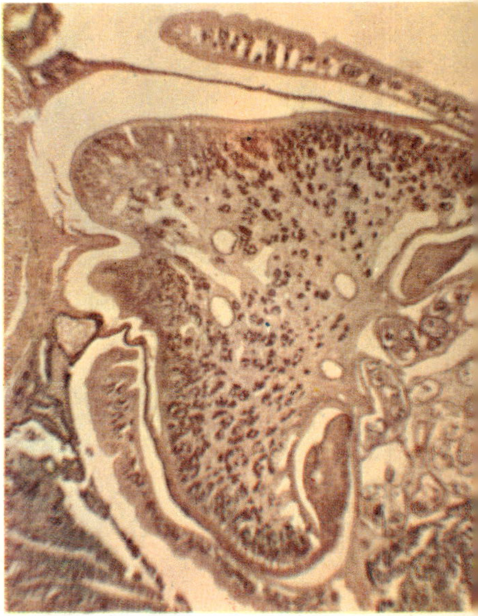

(d)

cestode is probably not a usual parasite of larids and is therefore eliminated when the physiological barrier of the adult bird becomes active. It would be interesting, in this connection, to know what kind of cestodes occurred in young cuckoos still confined to the nests of their passerine foster parents, since adult cuckoos contain only a single cestode endemic to the order Cuculiformes.

It might be expected that attempts to culture cestodes *in vitro* might have shed some light on these biochemical aspects of host specificity. Unfortunately, despite a considerable amount of work in this field and even a certain amount of success, the situation is still somewhat confused.

There is usually a marked immunological reaction when an adult or a larval parasite is closely associated with the tissues of the host (figure 8·2). At first sight, it might seem that helminths with larvae migrating in the host would be more likely to evoke such reactions than those which develop directly in the intestine. Yet many intestinal parasites burrow into the intestinal mucosa and may even destroy this. Among trematodes, for instance, strigeids destroy the mucosa with their tribocytic organ (figure 8·2) and are thus in direct contact with the blood capillaries of the submucosa. This is also the case for many cestodes which are deeply embedded in the mucosa and whose rostellar glands apparently destroy the epithelial cells (figure 8·2).

The cysticercoid of *Hymenolepis fraterna* develops within an intestinal villus of the mouse without there being any tissue reaction observable. In the blood of the mouse, however, there appear antibodies which prevent reinfestation and which also pass into the milk of lactating females, thus making the young mice resistant to infestation. These antibodies disappear after a certain time so that reinfestation becomes possible. When mice are infested with cysticercoids which have developed in an intermediate host, however, the antigen formed during development of the cysticercoid is absent so that reinfestation can occur. A host infested with larval trichina is resistant to reinfestation because the worms are unable to attach themselves to the intestinal mucosa due to the presence of antibodies resulting from the previous infestation.

There are, however, other phenomena which are probably not directly related to classical immunology. As long as a person harbours *Taenia saginata*, he or she may eat rare beef containing cysticeri without the latter growing into adult tapeworms. This is known as premunition for, as soon as the tapeworm is removed, infestation again becomes possible. In warm climates it is not uncommon to find persons with multiple infestations with this tapeworm, but these result from a single infestation with a large number of cysticerci since, in tropical countries, cattle are much more heavily infested than in our own. Premunition is probably also the explanation for cases of apparent immunity to malarial and leishmanial parasites. It is also very likely that premunition is not quite as simple as it seems since it does not appear when a host harbours tapeworms belonging to different species. For example, in the white mouse, *Hymenolepis microstoma* and *Catenotaenia pusilla* (figure 5·24) can occur simultaneously; in dogs, *Taenia pisiformis*, *T. serialis* and *Dipylidium caninum*. It is true that each of these species occupies a distinct region of the intestine, yet the size of the worms is not affected by a multiple infestation. There appears to exist a relationship between the size of the worms and the species of host or the size of its gut. *Diphyllobothrium latum* grows in man to a length of seven to ten metres, but in dogs and cats its length is only two metres and in hamsters the total length is only a few centimetres. In all cases the worms are adult and produce eggs but the width of the proglottids is relatively much smaller in cats and hamsters than in man. In multiple infestation in man with *D. latum* the worms are shorter and narrower than in single infestations, as witnessed by a case reported from Lausanne some fifty years ago in which eighty specimens were recovered, their length being about 60 cm and the width 3 mm. *Oochoristica incisa*, a tapeworm from the badger, usually occurs in large numbers, and here the length of the worms is about 10 mm. In single infestations, however, the worms grow to a length of 150 mm. This crowding effect is also known for other species of intestinal parasites, yet it cannot be related with certainty to immunological processes or to a lack of nourishment or even to both simultaneously.

Host specificity is due to both ecological and physiological factors and immunological effects must play some part. Worms are known to produce antigens consisting of proteins and polysaccharides, but their reactions to their hosts are difficult to interpret since this work must be carried out under artificial conditions which may not be relevant to those in the natural habitat. Also, the dose of antigen often used in order to stimulate the reaction is far greater than that which is likely to occur normally. For instance, it is most unlikely that a sheep on the pasture would swallow ten to fifteen thousand larvae of *Haemonchus contortus* in a single mouthful, which is the size of an experimental dose.

The pig *Ascaris* is unable to become adult in the human intestine. However, the larvae, which escape from accidentally swallowed eggs, migrate to the lungs but are later eliminated when they return to the intestine. Four distinct antigenic fractions have been isolated from pig *Ascaris* and are found to consist of glycoproteins. Two of these antigens can also be used to detect human ascarids. The presence of antigens common to the two species of ascarids explains the reason why larvae of pig *Ascaris* are able to sensitise human beings against a subsequent infestation. Yet this does not explain in what way larvae are prevented from growing into adult worms in the intestine. It is, moreover, curious to find that ascarid extracts also contain antigens of the blood groups A B O which are able to remove the agglutinins anti-A and anti-B from human serum and anti-A only from pig serum. The metabolic products of live *Ascaris* larvae, however, contain antigen A. Polysaccharide fractions containing antigens A and B have also been discovered in *Necator americanus*, *Trichinella spiralis*, *Taenia saginata*, and *Schistosoma mansoni*, but since the same antigens also occur in higher plants it is perhaps premature to attribute a role in host specificity to this 'molecular mimicry' (Damian, 1964).

Young birds still in the nest are very often heavily infested with intestinal worms whereas the parents are only slightly infested. The reason for this is that parent birds usually stuff their young with food, feeding them intermediate hosts containing larval forms. Once the young birds become independent they spontaneously get rid of their excess worm burden. This mechanism has been described in a young blackbird infested with *Porrocaecum ensicaudatum* in which the larval worms beneath the mucosa were surrounded by giant cells, an indication of an immunological reaction (Baer, 1961). Such self-regulatory mechanisms prevent a normal, healthy host from becoming too heavily infested, a situation unfavourable both for the parasite and for the host. The self cure reaction observed in nematodes belongs to this category and has been studied in detail in *Haemonchus contortus*, an intestinal parasite of sheep in which the larvae L_3 to L_5 are deeply embedded in the mucosa. It is clear that under such conditions it is the metabolic products of the larvae and especially the moulting fluid which are antigenic. When the course of a natural infestation is observed it is found that the amount of antibody present in the blood of a sheep is inversely proportional to the number of worm eggs eliminated, the maximum of the antibody curve corresponding with the minimum of the egg output curve. When large numbers of larvae are given experimentally it is found that they burrow deeply into the mucosa and that the amount of antibody in the blood increases. This phase is followed by the expulsion of the adult worms and the disappearance of eggs in the faeces. When the adult worms have been expelled the larvae, whose further development has been inhibited by the presence of antibodies, resume their normal development and grow into adult worms; eggs again appear in the faeces and the amount of antibody in the host's blood decreases. It is also possible to demonstrate the antigenic properties of larval metabolic products by placing larvae in serum with a high content of antibodies. Under these conditions a precipitate forms around the mouth, the excretory pore and the anus, thus indicating a specific reaction which is capable of inhibiting further development of the larvae. A similar reaction occurs when living cercariae of *Schistosoma mansoni* are introduced into the blood serum of a carrier. A mantle of precipitate is formed around each cercaria and immobilises it (Vogel and Minnig's test). In the case of *Nippostrongylus brasiliensis* (= *muris*) the life cycle in the rat includes larval migration through the lungs and antibodies appear in the host's blood. These antibodies prevent normal development when reinfestation occurs and larval development is slowed down and sexual maturity inhibited.

The specificity of antigenic reactions in worms is rather disappointing, however, especially when whole worms, either dried or mashed, are used as antigens. For instance, in the large liver fluke, *Fasciola hepatica*, antigens have been found which are also shared by *Schistosoma mansoni*, *Dicrocoelium dendriticum*, *Taenia saginata*, *Echinococcus echinococcus*, *Oncocerca volvulus* and *Trichinella spiralis*, not including the antigens already mentioned which act on human blood groups. Over twenty distinct antigen-antibody systems have been isolated from antigens prepared from *Schistosoma mansoni*. The antigen prepared from hydatid cysts and used for the Casoni test is not always specific, giving positive results in the absence of hydatids in man. One may hope that, with progress in biochemistry, the specificity of antigens can be increased in view of their usefulness for diagnostic purposes and for controlling the results of chemotherapeutical treatment in human and veterinary medicine. The example cited above of short-term self-vaccination observed in hosts harbouring *Hymenolepis fraterna* and *trichina* has led to the hope of obtaining specific antigens by using eggs or larvae previously irradiated by x-rays. Dried larvae and even adult worms of *Taenia taeniaeformis* injected into rats protect these from subsequent infestations. Yet the same method, using cysticerci of *Taenia saginata*, failed to protect calves from further infestation; so the amount of antigen produced might have been insufficient or it could have been destroyed. Such immunological reactions must be common among parasites in view of their selective advantage in preventing overinfestation. They may also explain the violent host reactions often observed in accidental hosts or in normal hosts weakened by disease or under stress, as is often the case with animals in captivity.

It may be possible in the future to envisage vaccination of domestic animals with synthetic specific antigens against their parasites so as to reduce the present considerable loss of food protein and perhaps also to preserve man from his parasite fauna. However, when the parasite burden of wild animals is compared with the higher burden in livestock, a biologist will conclude that man is responsible for this situation by keeping large herds on relatively confined pastures, a practice which, from a parasitological viewpoint, leads to an unnatural build-up of parasites.

These immunological reactions, though still imperfectly understood, occur in both intermediate and definitive hosts and must play a part in determining host specificity and in the formation of new races of parasites adapted to particular hosts.

On this basis it may seem rash to try and consider all the factors involved in the origin and nature of host specificity, but if the problem be approached from an ecological angle this might be made clearer. The host-parasite relationship must obviously have arisen in a biotope common to both host and parasite, and have had primarily ecological origins. Parasites would then have become isolated on certain hosts and this would lead to the development of parasite specialisation and physiological specificity. Monogeneans and nematodes may have evolved in this manner and in some cases seem to show a physiological specificity to their hosts which could have produced races adapted to particular hosts without showing obvious morphological modifications. The specificity of acanthocephalans and trematodes to their final hosts seems to be essentially of an ecological nature but the first intermediate mollusc host of trematodes exerts a filtering action on the trematode life cycle and therefore on its evolution. Even different races of the same species of mollusc may have different effects on the same parasite, whether they are present in the endemic area or not. Also, certain races of adult trematodes may possibly be adapted to particular definitive hosts; this may be true of strigeids. The selective action of the mollusc first intermediate hosts in trematode life cycles might be thought to favour development of a secondary physiological specificity from a basic ecological specificity. Examples from the life cycles of the frog parasites *Gorgoderina vitelliloba* and *Gorgodera euzeti*, however, warn against an over-'biochemical' interpretation of the facts. In *G. vitelliloba* the second host is a tadpole whilst in *G. euzeti* it is a *Sialis* larva. In both cases the cercariae are produced by the same species of bivalve. Figure 8·3 explains how it is the ecological conditions and negative geotropism of the cercariae of *G. vitelliloba* that produce this difference (Combes, 1968).

The nature of cestode specificity seems to be determined completely at the level of the final host. While ecological factors are important in bringing intermediate and definitive hosts into contact, it is in the end the taxonomic position of the latter that

8·3 Mode of infestation of the second intermediate host in two distinct species of trematodes but with the same bivalve as first intermediate host.

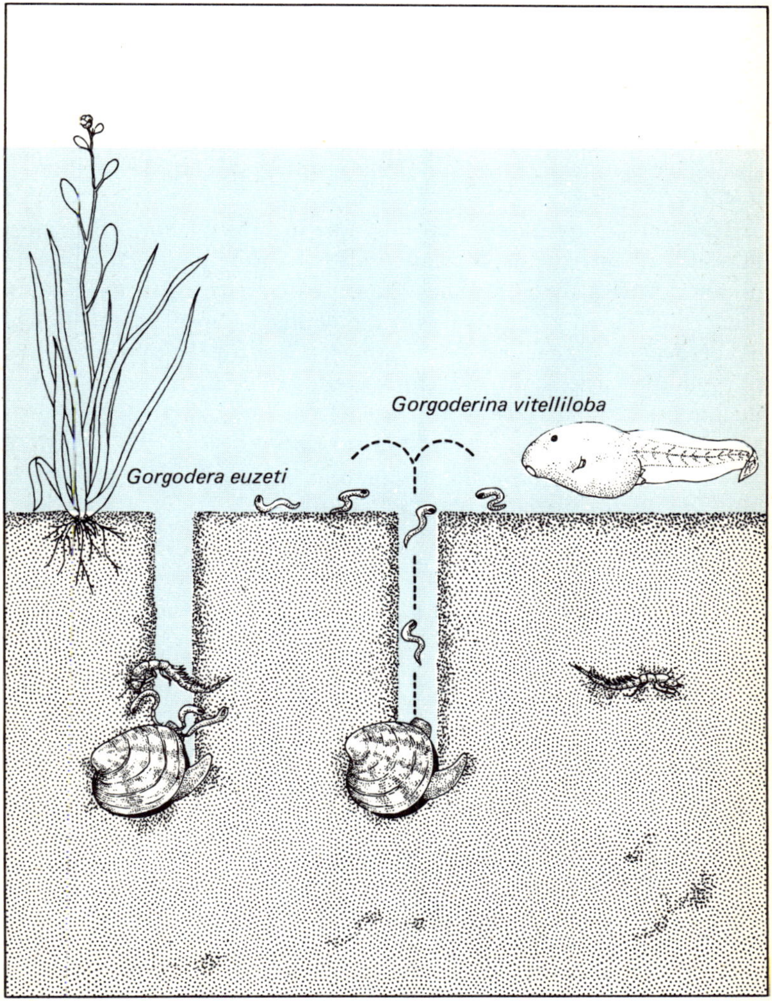

determines whether infestation may occur. As will be shown in chapter 9, a parallel evolution between cestodes and their hosts can be traced, which suggests that ancestral cestodes were already present in the ancestors of contemporary vertebrates, the two having evolved at the same pace. This is why the term phylogenetic specificity is appropriate to the relationships of cestodes with their hosts.

Thus host specificity depends on several factors, the most important being the ecological conditions bringing larval parasites in contact with their intermediate host or hosts, or in the case of direct life cycles, in contact with the definitive host. Establishment of the cycle in particular hosts can then lead to the appearance of physiological races of larvae, this then producing segregation into the definitive host. This would isolate parasites in particular hosts and create conditions favourable for speciation of the parasites. Thus the older the parasitic association the closer is the specificity until finally phylogenetic specificity results.

9 Parasites and evolution

The actual point in evolutionary history at which the parasitic habit developed in each group of proto-parasites has been a matter of some speculation. This problem can be considered from two angles, either from the point of view of evolution within a given class of parasites, or from the standpoint of the effect of host evolution on the parasites. This information can be projected back to give some idea of the kind of ancestral parasites that occurred on long-since extinct vertebrates, without actually having to resort to 'palaeoparasitology' since, except in the case of insects, there are no fossils to help here. Investigations of this kind stress the necessity of considering the host-parasite relationship as a biological unit with the host acting as the parasite environment. The fact that hosts and parasites have undergone a parallel evolution means that the evolution of the parasites can be used as a guide to the evolution of the hosts and can supplement morphological data about the latter. Similarly, host evolution can give clues as to the relationships of their parasites, for morphologically similar parasites often occur on related hosts.

Unfortunately it is not possible here to treat all groups of parasites in equal depth; in some cases because of too little information, while other groups are simply too small to allow extrapolations. This is true for the parasitic crustaceans and molluscs. The most interesting parasites from this point of view are those that show the greatest degree of host specificity.

Specificity in hippoboscids is usually largely ecological. *Hippobosca struthionis* occurs on ostriches but all the other species of this genus and even of the same subfamily live on mammals, both ruminants and carnivores. Ostriches share the same habitat as the plain-dwelling antelopes of Africa; furthermore, the texture of their skin is like leather and it is even used as such. Hippoboscids associated with mammals cannot, however, survive on birds so colonisation of ostriches must be due to convergence.

In general it seems that bird hippoboscids are the most primitive and have only secondarily become adapted to mammals. One would not expect to find, therefore, that the species parasitising mammals are more host specific than those living on birds. In actual fact, 75 per cent of mammalian hippoboscids are strictly specific while only about 20 per cent of those found on birds are host specific. It would be expected that since the birds arose in the Cretaceous, parallel evolution of parasites and hosts might have developed. The hippoboscids or their ancestors must have become obligatory parasites after the main orders of birds had arisen, while the mammals were still undergoing adaptive radiation. This would explain why forms which show little specificity to birds have relations that occur on related groups of mammals. The hippoboscids of birds are also less specialised in appearance. The most active period of evolution for the hippoboscids seems to have been in the Tertiary. As these evolutionary developments were occurring independently of the hosts it seems this was due to isolation of the parasite on the host; the parallel evolution of parasites and hosts would then have occurred more or less by accident.

The isolation of nycteribids and streblids on bats is the result of ecological factors and has led to specialisation on a par with that in hippoboscids, yet even more striking, due to the almost complete absence of wings, and also because of secondary factors such as the subcutaneous existence of the females of *Ascodipteron*. Bats are known from the Oligocene, yet most species alive today are much more recent. Here again the parasites of bats, although much more specialised morphologically than the hippoboscids, have also evolved independently of their hosts. Interspecific competition and other selection pressures appear to have hardly affected the streblids since a single species of bat can harbour eight species of parasite belonging to different genera.

With regard to the phthirapterans, it seems that the anopleurans

9·1 Presumed origin and present distribution of camelids. The circle indicates the probable region of origin.

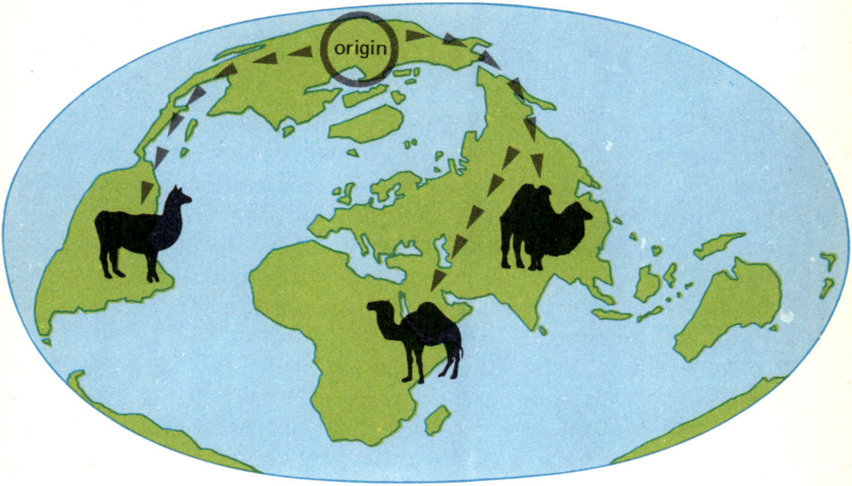

9·2 Below The distribution of horses for comparison with the distribution of ostriches in Africa and rheas in South America.

made an appearance on mammals at the beginning of the Tertiary, or perhaps even slightly earlier, at the end of the Cretaceous. The present distribution of lice shows pronounced specificity, especially in the more specialised forms. *Pediculus*, for instance, is specific to man and the chimpanzee but does not occur on any other primate, while the crab-louse lives only on man, chimpanzees and gorillas. Echinophthirians have already been mentioned and are exclusively parasitic on pinnipedes. These hosts, which arose in the late Miocene, harbour four parasite genera of which two occur on sea lions, one on seals and one on walruses, groups which are ecologically distinct from one another. *Echinophthirius* inhabits the nasal passages of seals, while the other genera live in the dense fur which retains air when the host dives. The sea lions are the most ancient group and harbour the genus *Proechinophthirius*, considered to be the most primitive parasite. Morphological features of pinnipede lice suggest that they are most closely related not to forms parasitising dogs, as might be expected, but to parasites of horses, pigs and large bovids (Hopkins, 1957). Thus these results shed no light at all on the origin of pinnipedes which, classed with fissipedes in the Carnivora, are placed in the Ferrungulata (Simpson, 1945) with the horses, suids and ruminants. In addition dogs, which are the only fissipedes having lice, are parasitised by a genus endemic to ruminants, suggesting that the canids acquired a secondary infestation after the pinnipedes had become aquatic.

Lice of the genus *Microthoracius* are specific to tylopods, a group also included in the ferrungulates. This genus probably arose in the Miocene on North American camelids and must have been transported on them, in the Pleistocene, into Eurasia and North Africa, where today they are represented by a form parasitic on the dromedary; they must have been taken into South America around this time, for here they occur on llamas (figure 9·1). Speciation, in this case, was probably due to ecological isolation and it seems likely that the three species of lice known today originally evolved on three species of llama, two of which have since become extinct. Thus of the three lice species known to occur on the llama, two must represent a secondary infestation (Hopkins, 1957).

The absence of anopleurans from certain hosts may be explained in several ways. For instance, no louse lives only on cetaceans, sirenids and on the hippopotamus, because of the hairless skin and aquatic way of life of these hosts. The absence of lice from Australian marsupials, which reached that continent at the end of the Cretaceous, can be explained by the fact that lice had not yet emerged at this time and were unable to colonise marsupials later as the continent of Australia had by this time become separate. On the other hand, the absence of lice and mallophagans from bats is probably related to the fact that the numerous pupiperans and fleas harboured by these mammals must have discouraged establishment of phthirapterans.

These few examples illustrate how the relationships of lice to their hosts are complicated by the fact that the specificity observed today is sometimes primary, sometimes the result of subsequent secondary evolution. Parallel evolution had already started before the major divisions of mammals had differentiated, so that this radiation must have eliminated many parasites, explaining why comparisons of parasite and host classifications can be confusing or even incomprehensible.

All orders of birds harbour mallophagans, but in the present state of taxonomy it is not possible to determine whether these infestations are primary or secondary. In general, forms living on the head are more specific than those occurring on the wings. Also, in cases where two groups of bird harbour different species of the same genus on their wings and body, mallophagans borne on the head belong to a separate genus. This could be a morpho-

9·3 Examples showing the ontogenetic development of various kinds of monogenean haptor, superimposed on a scheme showing the evolutionary relationships between their hosts (left half, larval haptor; right half, adult haptor).

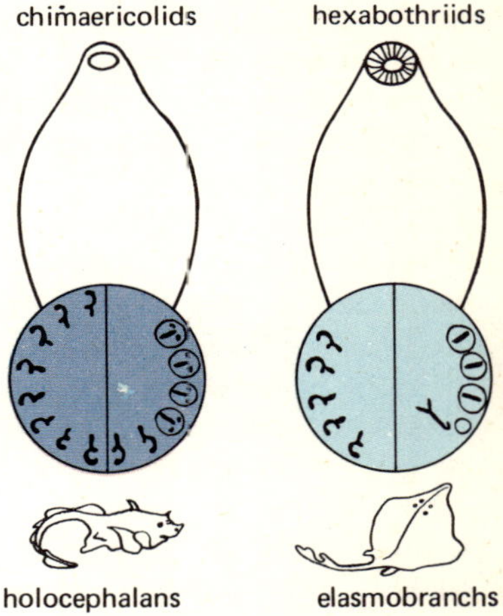

	Duration of Periods	Periods
Cenozoic	69	
Mesozoic	60	cretaceous
	35	jurassic
	35	triassic
Palaeozoic	30	permian
	20	pennsylvanian
	30	mississippian
	50	devonian
	30	silurian
	70	ordovician
	90	cambrian

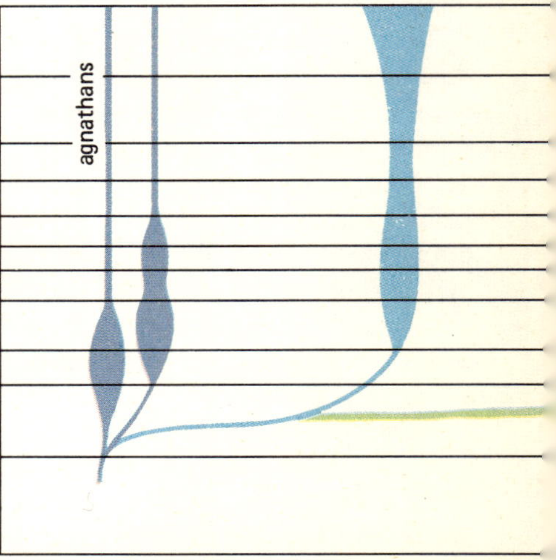

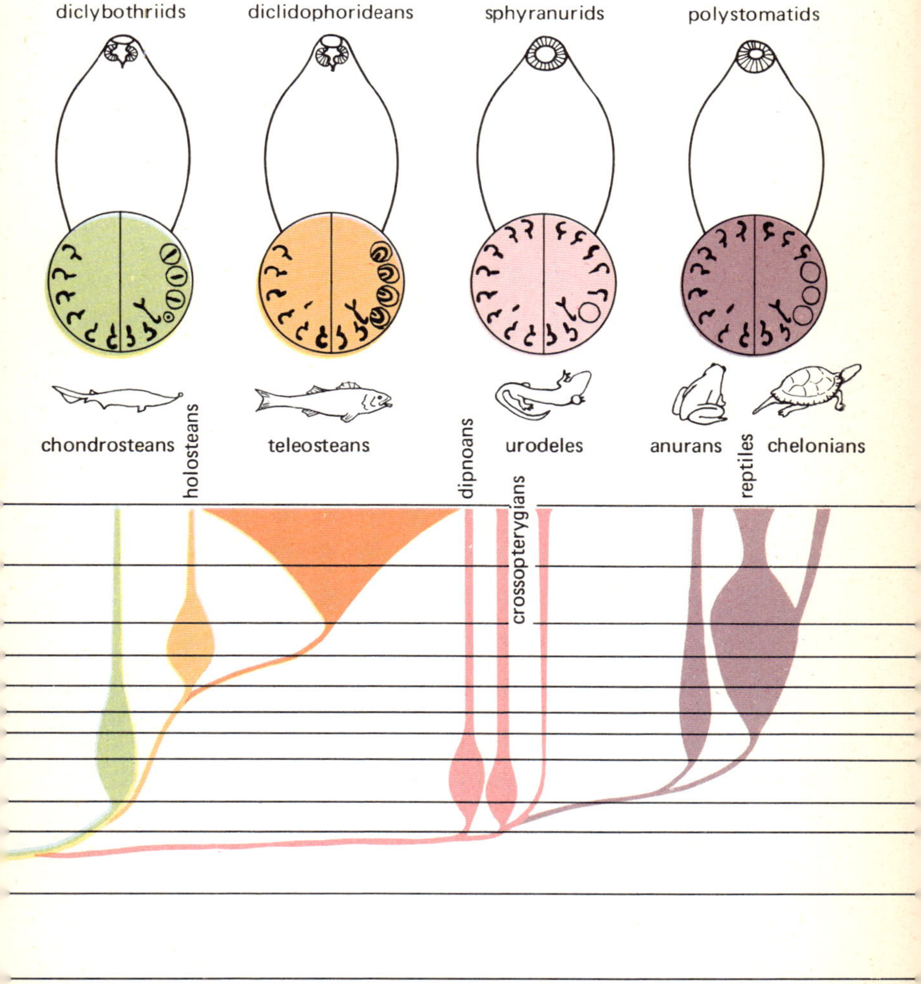
diclybothriids
diclidophorideans
sphyranurids
polystomatids
chondrosteans
holosteans
teleosteans
dipnoans
urodeles
anurans
reptiles
chelonians
crossopterygians

logical adaptation to short feathers, selection pressures finally having produced distinct generic characters. Although it is possible to find relationships between parasitised hosts, as with the anopleurans, these are not an expression of phylogenetic affinities and are usually due to similarities in feather structure. The case of ostriches and rheas parasitised by two species of the same mallophagan genus is difficult to interpret, since these birds have been geographically separated in South and equatorial Africa and South America respectively since the beginning of the Cretaceous. Although fossil remains of ostriches have been found as far away as China, there is no evidence that they occurred in North America and ever made contact with rheas in the Tertiary.

Both orders of bird inhabit dry savannah land and ostriches often occur with herds of antelopes and zebras. The rhea harbours *Struthiolipeurus* and the related genus *Meinertzhageniella*, while the ostrich harbours only *Struthiolipeurus*, suggesting that it became parasitised more recently than the rhea. It is tempting to suppose that the genus *Struthiolipeurus* was derived from mallophagans living on plain-living mammals such as equines. These were already established in North America at the end of the Miocene and migrated into Eurasia in the same period; migration into South America occurred at the beginning of the Quaternary and was followed by the extinction of horses in this continent. In Eurasia, horses would have continually met up with ostriches on their various migrations; for instance, all the way to South Africa where the zebras finally settled. The distribution of nematodes in these hosts also supports this idea because rheas and ostriches harbour the closely related genera *Deletrocephalus* and *Codiostomum*, which are classed with mammalian parasites, notably those of equines. Thus it seems that the presence of related mallophagans on rheas and ostriches is the result of secondary infestation in the Cenozoic and does not indicate an established phylogenetic relationship between these two orders of birds (figure 9·2).

Monogeneans are almost exclusively fish parasites and most authors now believe that fishes arose in fresh water in the Silurian about 400 million years ago and that the elasmobranchs split off from the ancestral stock in the Jurassic while the holocephalans, which are an even more ancient group, had already arisen in the Triassic (Marshall, 1957). The Chondrichthyes, which are almost exclusively marine, support a very varied monogenean fauna with families, subfamilies and genera that are absolutely host specific and which have arisen diphyletically, corresponding to the two major taxonomic divisions of monogeneans. Most of these forms live on the gills and in view of the high proportion of urea in the blood of chondrichthyans, must show physiological adaptations to their special diet here, an occurrence that would have tended to isolate them on this group of hosts at an early stage. This has resulted in the evolution of morphological differences, allowing three host-specific families and several host-specific genera and species to be identified, these probably having arisen as their hosts diversified into sharks, rays and holocephalans and became ecologically distinct. The monogenean fauna of modern representatives of this ancient stock cannot therefore be considered primitive. The parasitism must be fairly ancient, however, as is testified by the occurrence of related chimaericolid monogeneans on the gills of chimaerids and callorhynchids. These two holocephalans having probably already separated by the Cretaceous, are today confined to the seas of the northern and southern hemispheres respectively.

The evolution of bony fishes is more complicated because of their polyphyletic origins. One line, which arose in the Devonian and eventually produced contemporary Dipnoi and crossopterygians (*Latimeria*) may be free of monogeneans. The same is true of the holostean stock (*Amia*, *Lepisosteus*). In the chondrosteans the primitive polypterans are unparasitised, whilst the sturgeons do harbour monogeneans. The teleost stock has since the beginning of the Triassic undergone a remarkable adaptive radiation and today represents the most diverse group among the vertebrates (Marshall, 1957). It is possible that some day a monogenean may be found on the gills of the coelacanth (*Latimeria*), which is marine. The absence of monogeneans from most Dipnoi and *Amia* and *Lepisosteus* may be due to the fact that these fish are now found in fresh water while their ancestors were marine, the monogeneans having been unable to withstand the changeover. Of two species of *Diclybothrium*, however, one occurs on the gills of sturgeons while the other is found on the gills of *Polyodon*, the paddle fish, which is a native of fresh water in North America. The separation of the polyodontids and acipenserids occurred as far back as the Cretaceous and their monogenean parasites may well belong to a common stock which might have been either fresh water or marine. The fact that *Diclybothrium* is closely related to a group of monogeneans parasitising elasmobranchs suggests that they may be basically marine.

The marked host specificity of dactylogyrids and gyrodactylids parasitising fresh-water fishes has already been mentioned. Other fresh-water monogeneans are also specific, but to various degrees, which suggests that unrelated groups of host fishes may have undergone convergent evolution. In marine conditions, on the other hand, there seems to be a strict host specificity. Each species of gadid found in British waters harbours on its gills a species of *Diclidophora* peculiar to itself; *D. merlangi* on whiting, *D. luscae* on pout, *D. pollachius* on pollack, *D. minor* on *Gadus poutassou*, *D. denticulata* on the coal fish. The cod is not parasitised, but two other species of *Diclidophora* occur on *Urophycis blennoides* and *Macrurus rupestris* respectively. These fishes are not gadids but are related to them. Evolution on an isolated group of physiologically distinct hosts seems to have occurred here. The absence of *Diclido-*

phora from the cod is difficult to explain, but attempts to infest it with *D. denticulata* from the coal fish have proved unsuccessful. Further examples of this genus may come to light when a greater number of fish have been examined. In general, gill parasites are of much more use in tracing phylogenetic relationships between their hosts than skin parasites, where ecological conditions are often more important determinants of specificity than physiological factors. Thus the evolution of monogeneans and their hosts has not always proceeded in parallel and as a consequence the most archaic fishes do not harbour the most primitive skin parasites. The evolution of polyopisthocotylinean gill-parasitic monogeneans does, however, bear a close relationship to the evolution of their hosts, as Llewellyn (1963) has recently demonstrated (figure 9·3).

The cestodarians have in the past been considered to include two very different – probably in fact diphyletic – types, the gyrocotylids and amphilinids. The gyrocotylids occur in the spiral valve of holocephalans and their supposed life cycle (see p. 111) has led certain authors to propose that they are related to the monogeneans, though this could be a convergent feature. The chimaerids made their appearance in the Jurassic and were probably the only descendents of the bradyodonts, which arose in the Devonian and became extinct some 120 million years later in the Permian. The presence of *Gyrocotyle* in both *Chimaera* in the northern hemisphere and *Callorhyncha* in the southern hemisphere suggests that the ancestral holocephalans in the Cretaceous were already parasitised and that the parasites must have evolved at a slower pace than their hosts.

The other group of cestodarians, the amphilinids, inhabit the body cavities of various fish and, exceptionally, that of an Australian fresh-water tortoise. Unfortunately, the life cycle is known only for one species, *Amphilina foliacea*, a parasite of the European sturgeon, and this occurs in fresh water and involves a single intermediate host. Although it is not yet possible to relate the amphilinids to other cestode-like forms, their hosts belong to three major groups of fishes, of which the acipenseriformes are the most ancient while the clupeiformes and cypriniformes are the most primitive of the bony fishes. The fact that these hosts and their parasites are now spread over at least three continents also emphasises that the amphilinids are extremely ancient forms and probably extend back as far as the Jurassic.

Comparison of the classification generally accepted for cestodes with that of their hosts indicates that quite 'primitive', though specialised, hosts such as the elasmobranchs harbour very specialised parasites which are highly specific to them. The tetrarhynchs, tetraphyllids and diphyllids are quite morphologically distinct and have obviously evolved on different groups of hosts. The scolex of several forms bears surface sculpturing that relates to the particular pattern of villi or ridges on the intestinal mucosa (figure 9·4). It seems likely that sharks, rays and torpedoes come to harbour different cestodes because they have different feeding habits and consume different intermediate hosts. These cestodes must have differentiated very rapidly and it seems probable that they had already become specialised to elasmobranchs before these started to diversify at the beginning of the Cretaceous and had therefore probably already existed in the ancestral *Protoselachii*.

In view of the now generally accepted fact that the first vertebrates arose in fresh or brackish water it is particularly interesting that, of the few surviving holosteans, *Amia* is represented by two species, one of which occurs in fresh water in China, the other in fresh water in North America, and that the latter harbours a cestode, occupying a very isolated systematic position, which has four armed tentacles like a tetrarhynch and a strobila that fragments easily into independent proglottids armed with a pseudoscolex resembling that of pseudophyllideans (figure 9·5). It is

9·4 The scolex of *Monorygma perfectum*, which is adapted to the spiral valve mucosa of its host, a shark.

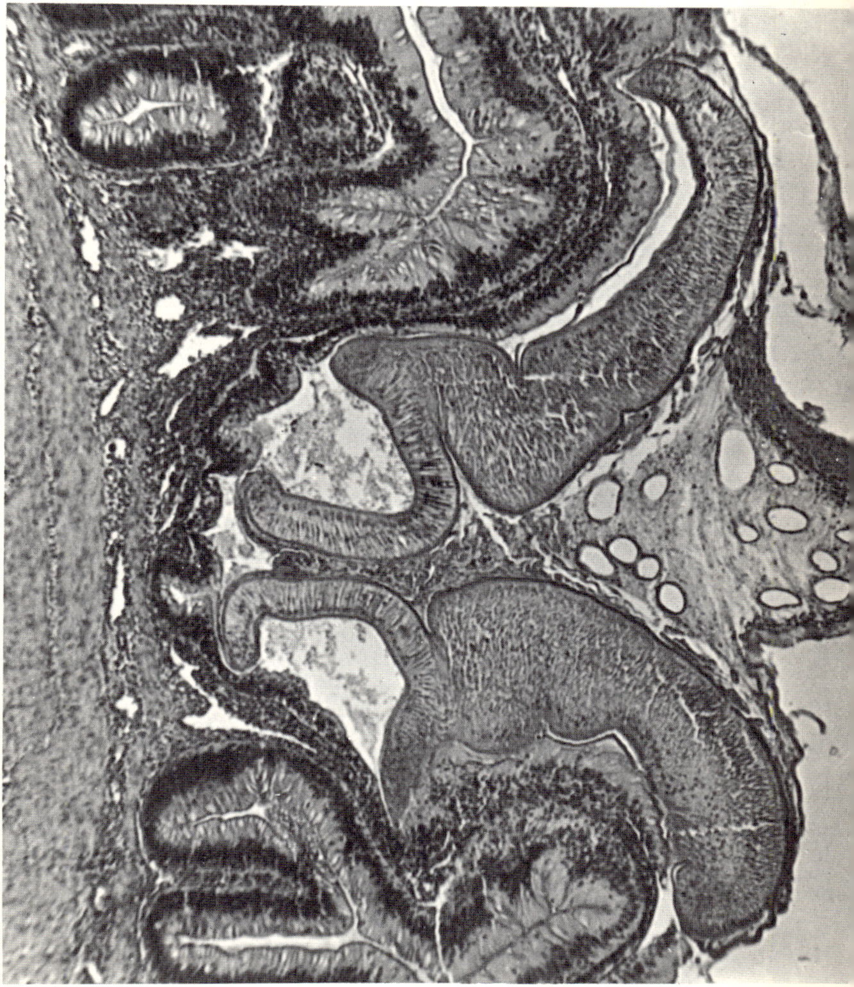

tempting to see this as a 'missing link' ancestral to the tetrarhynchs.

The cypriniformes, cyprinids and silurids are ancient and primitive fresh-water bony fishes. They harbour several subfamilies and genera of neotenic plerocercoids which occur in them throughout the world. The cestodes of marine teleosts are very different from those of fresh-water teleosts but are comparatively poorly known. Marine fishes harbour pseudophyllideans almost exclusively whereas fresh-water fishes harbour only a few pseudophyllideans but mostly ichthyotaeniids which have a scolex bearing four suckers and rather resemble the tetraphyllideans. The pseudophyllideans may have given rise to the line parasitising terrestrial vertebrates since there are examples of contemporary forms of this order in terrestrial vertebrates.

All groups of terrestrial vertebrates harbour cestodes, with the exception of crocodiles. The reason for this could perhaps be sought in the marine origin of modern crocodiles because it seems that marine hosts usually lose their parasites on passing into fresh water, either because their physiology changes or because the parasite life cycle becomes unable to function in the absence of suitable intermediate hosts.

Three closely related pseudophyllidean genera are harboured by varanids and pythons. They probably originated in the Cretaceous, since today varanids and pythons from Africa, Asia and Australasia harbour the same species of cestodes. In addition, palaeontologists consider that the boas, which are related to pythons, arose from a group of lizards almost identical to the varanids.

Without becoming involved in details of the distribution of cestodes in their bird and mammal hosts, one can say that each group of hosts below ordinal level and above the level of family has its own specific cestodes. It is impossible to infest a host experimentally with a larval cestode if it does not belong to the same group as the usual host. One cannot, for example, infest ducks with hen or pigeon cestodes, nor a rodent with a dog cestode. Man is exceptional in this respect and has secondarily acquired the parasites of his commensals, i.e. rodents and dogs, only two forms, *Taenia solium* and *T. saginata*, being morphologically and physiologically specific.

Man also harbours a species shared by monkeys (*Bertiella studeri*). It is often the case that hosts living in colonies such as rodents and conies or in flocks such as pigeons and guinea-fowl harbour cestodes which are extremely difficult to identify because they show considerable individual variation. This is due to the genetic constitution of cestodes, for these are often self-fertilising and give rise to clones, that is a number of individuals with the same genotype. When a number of hosts associate together in herds or flocks there is obviously a greater chance that each species of cestode will be represented by a number of clones. Physiological convergence of unrelated hosts which occupy the same biotope may have led to the transfer of two genera of cestodes from bustards to the ground hornbill and to some extent to the guinea-fowl; these three species of birds are savannah dwellers and share the same kind of diet (Baer, 1955). The intestinal environment is probably very stable, as is borne out by the fact that many genera are obviously direct mutations of known genera – for example those forms where the reproductive system is either partially or completely doubled. These mutants with replicated genitalia occur in the same orders of hosts as the cestodes from which they arose, suggesting that the mutation has not affected the course of the life cycle.

The host constitutes a very real ecological niche for the cestodes it supports, so that once the host becomes isolated speciation of its cestodes will occur. Flamingoes, for instance, are extremely specialised and difficult to relate to other aquatic birds. They harbour at least three equally specialised genera of cestodes peculiar to themselves. These have never been found to occur in ducks, which some authors consider to be related to flamingoes. Similarly stilts and avocets always support the same cestodes even when spread over different continents.

Adult trematodes occur in all the major groups of vertebrates, from the elasmobranchs to the birds and mammals, so one might expect that for many hosts the association must have been a long one. Manter (1957) has pointed out that of the relatively few species occurring in elasmobranchs none of these are primitively intestinal parasites. In fact it seems quite likely that most, if not all, elasmobranch digeneans have been secondarily acquired from teleost flukes and that as the teleosts arose in the Jurassic, or even later, parasitism of elasmobranchs must be more recent than this. Furthermore, these parasites have not evolved beyond the level of genus and have not formed distinct families. The trematodes of marine teleosts can be grouped into several families which are specific to a particular group of fishes, the plecognaths for example. These host groups are ecologically distinct, and therefore relatively isolated, from one another. The sun fish harbours seven species of the genus *Accacoelium* as well as two other monotypic genera belonging to the same family which are specific to it. Two other species of fish, unrelated to the sun fish but having the same ecology, also harbour related species of digeneans, however, while certain fishes actually related to the sun fish but differing from it in being bottom dwellers are not parasitised! In general, populations of marine fish living in similar conditions of density and temperature and at similar depths harbour the same trematodes.

In terrestrial vertebrates ecological isolation does not tend to be as marked and occurs within a given geographical region. African and Asian crocodiles, for instance, have different trematodes from New World crocodiles which harbour a family of diplostomatids specific to themselves. The Mexican crocodile belongs to the same genus as the African and Asian forms but is parasitised by diplostomatids specific to other South American crocodiles which are absent from Africa and Asia.

The trematode *Achillurbainia* was discovered in the cranial sinuses of two black panthers autopsied at the zoo in Vincennes (France) and was also found in the sinuses of a South American opossum (Dollfus & Nouvel, 1957). A related form had previously been recorded from postauricular abscesses in man in Kasai (Congo), Nigeria and the Cameroons. This strange geographical distribution can only be explained by assuming that the normal hosts are the great cats which spread from Asia (Panther) into Africa (Leopard) and then migrated to South America in the Pleistocene (Jaguar). The metacercaria-bearing intermediate host is probably a fresh-water crab, and this would explain the infestation of opossums. Systematic examination of the cranial sinuses of leopards and jaguars is required to confirm this hypothesis.

In recent investigation Kasimov (1964) has examined the trematode fauna of over a thousand birds from the Soviet republic of Azerbaidjan. Eighteen orders of hosts were represented, of which 22 per cent were parasitised. It was noted that there was a direct relationship between parasite fauna and the host diet. This was particularly marked when comparing the fauna of terrestrial gallinaceous birds feeding on snails and insects with that of marsh-dwelling forms consuming aquatic crustaceans, worms and molluscs. Most of the birds that overwintered in this region had come from eastern Siberia, where they nested, and a comparison of the trematode fauna before and after hibernation showed that of sixty species, about half were common to both geographical regions while the other half were exclusive to the winter quarters. It is not known whether the latter species are lost on their northern migration or whether they fail to perpetuate themselves in the nesting grounds because of the absence of suitable intermediate hosts.

Thus although trematodes have often evolved within an isolated host or isolated groups of vertebrates, there are no examples to show that they have evolved in parallel with their hosts. If – as seems likely – digeneans were primitively parasites of molluscs, this group can be no older than the molluscs. Despite the enormous number of fossil molluscs known, it is generally agreed that the first gastropods appeared in the Lower Palaeozoic and that most forms had already become extinct by the Permian, only a few survivors persisting into the Upper Cretaceous or beginning of the Tertiary. These primitive forms consisted of terrestrial forms which then reverted to an aquatic existence and passed back into the sea via fresh water. Most contemporary gastropods, however, arose towards the end of the Cretaceous and at the beginning of the Tertiary. Although a large number of trematode life cycles still have to be worked out it is nevertheless possible to see that comparatively few of them use a lamellibranch or scaphopod as first intermediate host and annelids are used only exceptionally. The great majority involve gastropod first intermediate hosts. The colonisation of fishes by trematodes must therefore be relatively recent and must have occurred when the teleosts were starting to undergo the explosive adaptive radiation that allowed them to occupy many aquatic ecological niches where they became even more specialised. The evolution of a metacercaria that could develop directly into an adult worm without undergoing a complicated metamorphosis must certainly have favoured the invasion of terrestrial vertebrates, within which trematodes are again distributed according to ecological factors rather than host-phylogenetic relationships.

It is extremely difficult to gain a general view of host specificity among nematodes since their polyphyletic origin implies that they adopted parasitism independently and at different times during the evolution of their hosts. It follows that forms parasitising mammals and birds are not necessarily more 'highly evolved' than the parasites of fishes or amphibians. Major phyletic lines following an overall evolutionary direction are rare or may be more apparent than real.

An evolutionary progression, however, has been traced for ascarids, which extend from the elasmobranchs, through the teleosts, amphibians and reptiles into the birds and mammals (Osche, 1958). When one considers the age of the hosts which span the 430 million years between the Palaeozoic and the Pleistocene, and looks for parallel evolution in their nematode parasites, it is found that the latter have only reached the level of super-family in this time. Even if it is assumed that parasite evolution has occurred much more slowly than that of the hosts, it would be expected that a greater number of species and genera would occur in the elasmobranchs than the two genera with few species that actually occur here, even though these constitute a separate family. The ascarids probably extend back as far as the Cretaceous as is suggested, for instance, by the fact that they occur in both Old World and New World crocodiles. During the Cretaceous, ascarids apparently transferred from lizards to the snakes which evolved from them and are today almost completely absent from lizards; the species *Ophidascaris filaria* occurs in the pythons of Africa, Asia and Australasia (Sprent, 1969). Crocodiles and tortoises also harbour ascarids. A similar situation also exists among the mammals, where primitive forms like the insectivores are not parasitised whereas carnivores are. The life cycles of most ascarids, where known, are nearly always heteroxenous, usually with two intermediate hosts. Among the mammals the acquisition of ascarids by herbivores and primates is more recent and the fact that the life cycle has become secondarily monoxenous in these forms is related to this recent parasitism (see p. 98).

Despite the difficulty in tracing lines of nematodes back to the

9·5 Scolex of *Haplobothrium*. (**a, b**) primary scolex; (**c**) secondary scolex.

free-living forms from which they arose, and of establishing the existence of phylogenetic specificity, there are examples of related nematodes that have evolved within members of the same or related groups of hosts. Dougherty (1951) has made a detailed study of metastrongyles in this respect. These highly specialised worms live in the lungs and circulatory system of their host and have a heteroxenous life cycle. The origins of this group can be traced back to the Cretaceous since primitive forms occur not only in marsupials, carnivores and suids but also in insectivores, so they could have arisen either before the separation of the mammals and marsupials or before formation of the Australian continent. Three other lines lead to carnivores, artiodactyls and whales. Of the latter, only the odontocetes support metastrongyles, but as the pinnipedes are also parasitised it is easier to imagine that metastrongyles were passed from terrestrial carnivores via them to cetaceans, than to look for a common ancestral form which would have been shared by whales, pinnipedes and artiodactyls.

The oxyurids are probably the oldest and most distinct group of nematodes. They are parasites of both invertebrates and vertebrates, which is fairly exceptional amongst nematodes, and among vertebrates occur primarily in amphibians, reptiles and, in mammals, in rodents and primates. At first sight if might seem that this distribution is associated with the herbivorous habits of the hosts involved rather than because of a long evolutionary association, and the secondary establishment of an oxyurid in one marsupial is certainly due to host diet. This group is extremely old, however, as is shown by the fact that tortoises of the Galapagos Islands, of Africa, Asia and the Indian peninsula, all harbour the same genera of oxyurids, so this does suggest that their ancestors were already parasitised by oxyurids at the end of the Jurassic or in the Cretaceous. *Enterobius* occurs in both New World and Old World monkeys and has probably been acquired by man only

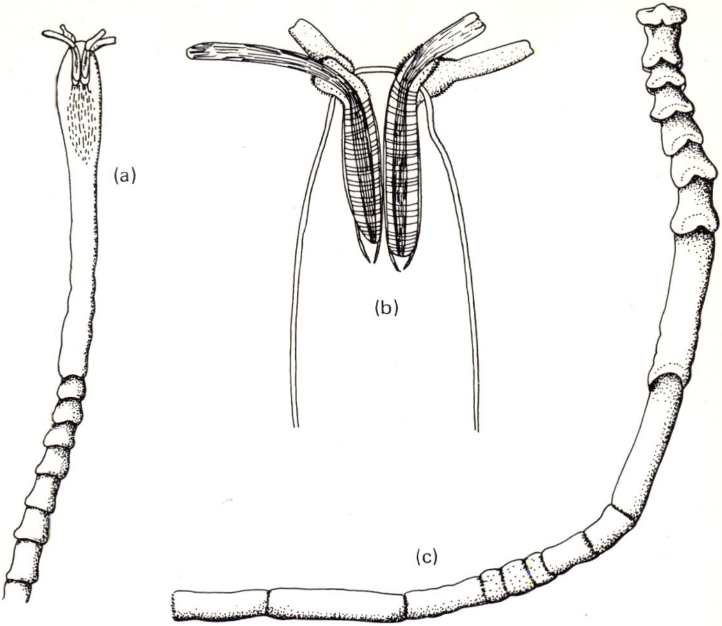

comparatively recently; the first African hominids may have contracted the infestation from monkeys that associated with them and their dwellings.

The Acanthocephala offer few clues as to their evolutionary history, although their comparative lack of host specificity points to the fact that they may have taken up parasitism fairly recently. Acanthocephalans are extremely specialised worms which cannot easily be related to any other group of invertebrates; there is some evidence, from the occurrence of *Oncicola* in Old World and New World cats, that these parasites arose at the beginning of the Tertiary, towards the end of the Eocene. The occurrence of *Prosthenorchis* in Madagascan lemurs and monkeys of South America, Asia and Africa can also be cited in evidence of the above dating. These examples are, however, exceptional, and elsewhere specificity seems to be of an ecological and physiological nature.

Larval acanthocephalans show a marked ability to re-encyst should an unfavourable host be encountered, which explains the extremely large number of paratenic hosts that have been recorded (Golvan, 1959–62). While some of these have become essential to the life cycle (see p. 168), others represent a dead end.

This chapter has set out to examine the possibility that parallel evolution between parasites and their hosts may have occurred; this would suppose that parasitism was acquired very early by many groups and that once this was established the parasite would evolve together with the host, but at a slower pace. The monogeneans, cestodes and nematodes are extremely ancient groups which extend back almost as far as the Tertiary and occur today in association with the oldest living vertebrates, some of which first appeared in the Jurassic. The monogeneans and cestodes seem to have evolved more rapidly than the nematodes and unlike the latter have differentiated into several distinct lines. In the case of elasmobranch cestodes, several evolutionary lines can be recognised on the basis of sculpturings on the suckers, which have become adapted to the very variable patterns on the mucosa of the host's spiral valve. In teleosts, where the structure of the intestinal mucosa is more conservative, the scolex of their cestodes is simplified and usually represented by four small suckers. The same kind of scolex characterises the cestodes of terrestrial vertebrates, but is usually supplemented with a rostrum armed with a circlet or circlets of hooks. Although the scolex of tapworms from terrestrial vertebrates has a remarkably constant structure throughout the group, other features are much more variable and this is especially true of tapeworms in birds. The major evolutionary lines of cestodes can be superimposed on the evolutionary scheme for the vertebrate hosts, for it is apparent that the primitive vertebrates harbour the most primitive cestodes while the more 'recent' hosts, in particular the birds, harbour the most 'highly evolved' cestodes.

Thus the long-standing relationship between cestodes and their vertebrate hosts led to phylogenetic specificity and parallel evolution between the two (Baer, 1946). Because, as seems likely, the cestodes have evolved in concert with their hosts, but at a slower pace, it should be possible to use phylogenetic relationships between the parasites to establish evolutionary relationships between the hosts.

The tetrahynchs, which are characterised by a scolex bearing four protrusible tentacles armed with hooks, are specific to elasmobranchs. A similar kind of scolex, though more developed, is a feature of a form parasitising *Amia*, a North American fresh-water fish. This host, known only since the Cretaceous in fresh-water deposits, is descended from a long line of fossil fishes which extend back as far as the Permian and were contemporary with the first elasmobranchs. Judging from their cestodes, the relationships between *Amia* and elasmobranchs might be closer than has been appreciated.

Four orders of sea-going birds harbour specialised cestodes which, although the life cycle is not known, may use cephalopods as intermediate hosts. When evolutionary lines are established for the parasites using anatomical features, it is found that these coincide with the classification of the birds (Baer, 1954). The South American bird *Cariama* has been classified among the cranes, where it occupied a rather isolated position, but was found to harbour two specialised cestode genera which had otherwise been found only in Eurasian and African bustards. The fact that *Cariama* possessed cestodes (and nematodes) related to those of bustards confirmed the opinion of certain ornithologists that in fact this bird was more closely related to the bustards than to the cranes.

It is more difficult to establish phylogenetic relationships between mammals, since many orders disappeared in the Tertiary together with their cestodes. The position of conies (Hyracoides) is somewhat problematical. They are ungulates with a dentition similar to that of the rhinoceros, but in size and other respects they resemble rodents. At least two families and several fossil genera are known and are confined to Africa and the eastern Mediterranean coast, a distribution corresponding with that of contemporary forms. The cestodes of hyraxes show affinities with those of both ungulates (perissodactyles) and rodents (murids): the genus *Anoplocephala* is shared with zebras and the rhinoceros, and the genus *Inermicapsifer* with African murids.

10 Conclusions

Although the scope of this book has not allowed all the various kinds of parasites to be dealt with, an attempt has been made to select those features most characteristic of parasites and also to give examples of the specialisations associated with a parasitic way of life.

Parasites are distinguished from free-living forms in that they live in external or internal association with other living organisms. The parasite's ecological niche is formed by the host so that the parasite is not tied to any particular biotope, for this may be changed by the host in response to seasonal or other environmental influences or due to endogenous stimuli. Migratory birds, for instance, may carry their parasites many miles away from the site of original infestation. It seems likely that the mobility of the host has exerted a selective influence on parasites and in particular on the way in which the life cycle is completed, especially where intermediate hosts are involved.

Most of the parasites mentioned in the preceding pages arose towards the end of the Cretaceous or at the beginning of the Tertiary. Some of them, such as the hippoboscids, lice and some nematodes, probably arose more recently while others, like the cestodes, extend right back to the Palaeozoic.

Gastropod molluscs are known to have originated in the Palaeozoic but most of them had already disappeared by the Permian. Their survivors, which reached the end of the Cretaceous and beginning of the Tertiary, consisted essentially of terrestrial forms, several of which became secondarily aquatic. Towards the end of this period appeared the forms which have survived until today. It seems quite likely then that the trematodes also originated at the beginning of the Tertiary.

Parasite evolution has therefore been taking place for at least seventy million years, during which time the host-parasite relationship has become established. There is every reason to believe that when hosts became extinct many of their parasites became extinct with them, also that many parasites must have perished by natural selection. The sudden climatic changes during the Tertiary, followed by the Ice Ages of the Pleistocene, produced profound modifications in ecosystems and their fauna; they certainly affected intermediate hosts unfavourably and must have affected many parasite life cycles. When these kinds of factors are considered the effects of positive selection pressures favouring increased fecundity of parasites and larval multiplication become much more significant. The marine environment, which has to a large degree escaped such fluctuations, has tended to preserve its parasite fauna so that the evolution of the parasites and their life cycles has proceeded in a more or less unhampered fashion, as witnessed by the monogeneans and especially by the trematodes and cestodes.

Host specificity is a concomitant of this extended period of adaptation and reflects the kind of ecological conditions in which it arose. Each group of parasites has managed to satisfy the physiological requirements necessary to its development, at the same time becoming adapted to life on particular hosts.

A truly objective analysis of the kind of adaptations that occur in relation to a parasitic way of life is not easy as it is all too tempting to resort to anthropocentric arguments which can only be offset by experimental observation. In some groups, like the nematodes, such speculation is almost impossible owing to their considerable degree of preadaptation to parasitism, also because of the polyphyletic origins of parasitism in this group. It is more than likely that the initial adaptation was physiological and that this was followed by morphological changes once the nematodes started to specialise and become isolated on a particular host or group of hosts. The life cycle would also become specialised, becoming associated in some forms with migration within the definitive host or with an increasing isolation of larval stages into intermediate hosts. The increased size of nematodes, especially of females, would augment the number of eggs produced, which is much greater than in free-living nematodes. In regions where human filariasis is endemic, it is quite common to capture infested insect intermediate hosts which demonstrates how enormous the production of microfilariae accumulating in the peripheral blood of the host must be.

Parasitism in molluscs and crustaceans is always associated with anatomical modifications which are more consequences than causes of a parasitic way of life. The atrophy and reduction of some organs which have become superfluous is often quite striking; thus locomotory appendages are often lost from crustaceans and the foot reduced in parasitic molluscs; on the other hand these organs may take on new functions, transforming into attachment organs or becoming a brood chamber for eggs, like the foot of molluscs. In parasitic molluscs the eggs are never laid directly into sea water but are incubated in a pocket developed from an existing organ which may be different in different groups. This brood chamber represents the main adaptation to parasitism in molluscs and enables late infestive veliger stages to be released which will find little difficulty in locating a host in the vicinity since the echinoderm hosts generally live in groups.

Parasitic crustaceans characteristically produce a great number of eggs, and in the epicaridians eggs are incubated in a brood chamber so enormous that these forms have lost all obvious resemblance to crustaceans. The rhizocephalans, which can be recognised as crustaceans by their larval stages, are the most specialised of the parasitic crustacea, having a unique form of larval penetration into the host which must be one of the most astonishing adaptations to parasitism known.

Dwarf neotenic males occur in both molluscs and crustaceans; they only very rarely occur in free-living members of these groups, for instance in a few cirripedes. It is possible that the ancestors of these parasites were hermaphrodites because protandric hermaphroditism occurs, for example in the cryptoniscids. The presence of a single female on a host is sufficient to cause any males to become neotenic and reach sexual maturity while retaining larval characters. Each host therefore carries only one female parasite and it is tempting to suppose that this might be a way of conserving the resources of a host which could not support several normally sized parasites. In the sacculinids, which were once believed to be hermaphrodite, the neotenic males which develop no further than the cypris stage are already determined in the egg, so there are male and female eggs distinguishable only by size. The parasitism of sacculinids would not be able to survive if the crab hosts lived independently of one another, as all the eggs produced by a single female are of the same sex; also the neotenic males, lacking a gut, are short-lived and would not survive during the time it takes the internal *Sacculina* to become the mature external form. This does not appear to jeopardise the success of the parasite to any serious degree as it is compensated for by the behaviour of the crab host; in some littoral regions the proportion of parasitised hosts is very high while in others there is a low infestation rate.

Unfortunately the biology of parasitic molluscs is poorly understood and the origin of the neotenic males is not really known. In some endoparasitic forms which have lost contact with the exterior the neotenic males which are very small have been mistaken for the testis; so these forms are physiologically hermaphrodite. The protelan parasitism of unionids and mutelids is quite difficult to understand, since no other fresh-water or marine lamellibranchs really show similar tendencies. In the fresh-water sphaerid bivalves, however, the eggs are incubated in special sacs where the embryos are nourished by a kind of placenta and where development to an advanced stage occurs before the eggs are laid. The rapid development that results is probably associated with the short life span of *Sphaerium* which, unlike other bivalves, lives no longer than a year. The unionids, on the other hand, develop slowly and only become sexually mature after four years, continuing to grow for several years. With regard to other lamellibranchs, the edible mussel produces about twelve million eggs, which are shed into the sea to develop, while oysters, which incubate their eggs, produce only two million larvae. The eggs or larvae of these forms develop without metamorphosis. The unionids and mutelids are the only lamellibranchs where eggs incubated by the females produce larvae which have had to become obligatory parasites of fishes in order to be able to complete their metamorphosis. Transitory larval stages such as glochidia and lasidia are unknown in other lamellibranchs, even those of fresh water. These larvae are equipped with attachment organs facilitating their fixation to the host fish. Unionids are known to have existed in fresh water since Silurian times and were widespread in the Carboniferous, an age when the fishes started to proliferate. This kind of parasitism must be very ancient and would have had to adapt to different fish hosts when the first hosts became extinct until these bivalves came to occupy the ecological niches they are found in today.

The increased fecundity of platyhelminths has been brought about in two ways. In trematodes a large number of cercariae are produced from a single egg by polyembryony, while in cestodes continuous growth of the strobila and uninterrupted maturation of the segments ensures a constant supply of a large number of eggs. In addition to this primary process in cestodes, larval multiplication has been developed and can attain staggering proportions, as in *Echinococcus*. It is, however, interesting that adult *Echinococcus* in the gut of the dog consists only of three to five segments and that comparatively few eggs are formed in the gravid uterus which occupies only the last segment. This 'principle of compensation' occurs in two other tapeworms in shrews. Adult *Hymenolepis prolifer*, in the common shrew is merely 50 mm long and the uterus only contains about a dozen eggs, but the larval form buds off hundreds of cysticercoids from a kind of plasmodial mass (figure 5·21). The white-toothed shrew also harbours a small form, *Hymenolepis pistillum*, which undergoes larval multiplication. In these cases it seems likely that larval reproduction compensates for the small size of the adult worms, which can produce only a limited number of eggs. This process would also favour survival of parasites in hosts which have an average life span of only a few months.

The metacercariae of trematodes and the plerocercoids and cysticerci of cestodes represent resting stages in the life cycle which are infestive and await the arrival of a suitable definitive host. Such larval forms can survive for quite long periods, especially where they occur in vertebrates, in a fish or a mammal. They may remain infestive for several years, the length of time depending on the time at which infestation occurred and on the longevity of the intermediate host. Metacercariae and cysticercoids which occur in invertebrates stand a much greater chance of perishing before the life cycle can be completed, due to the comparatively short life span of such hosts.

Ducks, inhabiting shallow Icelandic lakes, are parasitised by trematodes and cestodes, and during their autumn migration transport their parasites away from their intermediate hosts. During winter the lakes freeze right through and the intermediate hosts are killed off by the cold. Only intermediate hosts infested in spring, following the return of the migrants, are capable of infesting the new generation of ducks. This shows how a sudden change in an ecosystem could eliminate a large proportion of parasites by destroying the intermediate hosts and applies equally well to life cycles using terrestrial arthropod and mollusc intermediate hosts, which tend in general to be longer lived.

No parasite, whether adult or larval, can survive outside its host. The latter represents the controlled ecosystem where the parasite lives, buffered from external conditions, rather like the isolated fauna of a floating island. This explains how it is that conditions are so favourable to parasite specialisation and how it is that in self-fertilising hermaphrodites, such as most trematodes and cestodes, clones of genetically homozygous individuals are produced. In this kind of situation even recessive mutations will be expressed in the phenotype almost immediately since the mutation will occur rapidly in both the eggs and sperm of the hermaphrodite and a homozygous individual will be produced which may then give rise to a clone of genetically identical individuals by polyembryony, all manifesting the mutant character (see p. 144). It is easy to imagine how such clones could become rapidly adapted to 'new' strains of hosts occurring in the same ecosystem.

The more specialised and confined the ecosystem, the less chance there is that mixing of the gene pool will occur during the life cycle, but in less specialised conditions genetic recombination will be considerable and may produce a marked morphological instability in the parasites. Physiological host specificty and, in the case of the cestodes, phylogenetic specificity, do however impart a certain degree of stability to this system.

To summarise, parasites have a very special relationship with regard to their host environment, quite unlike that experienced by any free-living animal relative to its physical milieu. The host can also transport the parasite into quite different ecosystems where the parasite life cycle may be unable to complete itself. But the considerable reproductive potential of parasites allows them to maintain themselves, and in some cases if a host dies transfer to another host of the same species, or even to an unrelated host, allowing survival.

Thus parasites are subject to the same kinds of laws as all other living things with the reservation that the parasitic existence necessarily requires relationship with a host organism which constitutes a specialised ecosystem. Parasites are the only living things to occupy a microhabitat independent of changes in the surrounding macrohabitat; as a result they escape, to some extent, interspecific competition, a factor which may explain the comparative slowness of their evolution relative to that of their hosts.

Bibliography

General works

Jean G. Baer, *Le Parasitisme*, Paris, 1946, *The Ecology of Animal Parasites*, Urbana (Ill.), 1951; M. Caullery, *Parasitism and Symbiosis*, London and New York, 1952; N. A. Croll, *Ecology of Parasites*, London and New York, 1966; V. Dogiel, *General Parasitology*, Edinburgh and New York, 1964; P. P. Grasse, *Parasites et Parasitisme*, Paris, 1935; E. R. and G. A. Noble, *Parasitology: The Biology of Animal Parasites*, London and Philadelphia, 1961; M. Rothschild and T. Clay, *Fleas, Flukes and Cuckoos*, London and New York, 1952; J. D. Smyth, *Introduction to Animal Parasites*, London and New York, 1962; E. J. L. Soulsby (ed.), *The Biology of Parasites: Emphasis on Veterinary Parasites*, London and New York, 1966.

1 The parasitic way of life

M. S. Henry, *Symbiosis*, London and New York, 1966–7; J. F. A. Sprent, *Parasitism*, London and New York, 1963.

2 Adaptions to parasitism

C. Boquet and J. H. Stock, 'Some Recent Trends in Work on Parasitic Copepods', *Oceanogr. Mar. Biol. Ann. Rev.*, **1**, 289, 1963; V. A. Dogiel, *General Protozoology* (2nd ed.), Oxford and New York, 1965; R. Elsdon-Dew, 'The Epidemiology of Amoebiasis', *Adv. Parasit.*, **6**, 1, 1968; V. Fretter and A. Graham, *British Prosobranch Molluscs: Their Functional Anatomy and Ecology*, London and New York, 1962; G. Fryer, 'The Developmental History of *Mutela bourguinati* (Ancey) Bourginat (Molusca: Bivalvia)', *Phil. Trans. R. Soc. London, Ser. B.*, **244**, 259, 1961; P. C. C. Garnham, 'Malaria in Mammals Excluding Man', *Adv. Parasit.*, **5**, 139, 1967; K. G. Grell, *Protozoology*, New York, 1956; C. A. Hoare, 'Evolutionary Trends in Mammalian Trypanosomes', *Adv. Parasit.*, **5**, 47, 1967; A. Ichikawa and R. Yanagimachi, 'Studies on the Sexual Organisation of the *Rhizocephela* I.', *Annot. Zool. Jap.*, **31**, 82, 1958; **33**, 42, 1960; J. Lützen, 'Unisexuality in the Parasitic Family *Entoconchidae* (Gastropoda: Prosobranchia)', *Malacologia*, **7**, 7, 1968; E. G. Reinhard, 'Experiments on the Determination and Differentiation of Sex in the Bopyrid *Stegophryxus hyptius* Thompson', *Biol. Bull.*, **96**, 17, 1949; R. Yanagimachi, 'Studies on the Sexual Organisation of the *Rhizocephala* III', *Biol. Bull.*, **120**, 272, 1961.

3 Ectoparasitic insects

J. C. Bequaert, 'The Hippoboscidae or Louse Flies (Diptera) of Mammals and Birds', *Entomol. Americana*, **32**, 1, 1953; **33**, 1, 1954; G. H. E. Hopkins, 'Host Associations of Siphonaptera', *1er symp. spéc. paras parasites de Vertébrés, Neuchâtel*, 64, 1957; O. Theodor, 'Parasitic Adaptation and Host-Parasite Specificity in Pupiparous Diptera', *1er symp. spéc. paras. parasites de Vertébrés, Neuchâtel*, 50, 1957.

4 Round worms or nematodes

J. E. Alicata, 'Biology and Distribution of the Rat Lungworm, *Angiostrongylus cantonensis*, and its relationship to Eosinophilic Meningoencephalitis and Other Neurological Disorders of Man and Animals', *Adv. Parasit.*, **3**, 223, 1965; J. Robinson, D. Poynter and R. J. Terry, 'The Role of the Fungus *Pilobolus* in the Spread of the Infective Larvae of *Dictyocaulus viviparus*', *Parasitology*, **52**, 17, 1962.

5 Flat worms or platyhelminths

G. C. Kearn, 'The Life Cycles and Larval Development of some Acanthocotylids (Monogenea) from Plymouth Rays', *Parasitology*, **57**, 157, 1967; H. W. Manter, 'Studies on *Gyrocotyle rugosa* Diesing, 1850. A Cestodarian Parasite of the Elephant Fish, *Callorhynchus milii*', *Zool. Pub. Victoria Univ. Coll.*, 17, 1951.

6 Acanthocephalans or thorny-headed worms

Jean G. Baer, 'Embranchement des Acanthocéphales', *Traité de Zoologie*, **IV**, 733, 1961; Y. Golvan, 'Le Phylum Acanthocephala', *Ann. Parasit.*, **33**, 538, 1958; **34**(1), 138, 1959; **37**, 1, 1962.

7 How parasites infest their hosts

Jean G. Baer, 'Host Reaction in Young Birds to Naturally Occurring Superinfestations with *Porrocaecum ensicaudatum*', *J. Helminth. Supp.*, 1, 1961; R. M. Cable, 'Marine Cercariae of Puerto Rico', *Sc. Survey of Porto Rico and the Virgin Islands*, **16**, 491, *New York Acad. Sc., New York*, 1956; G. H. E. Hopkins, 'The Distribution of Phthiraptera on Mammals', *1er symp. spéc. paras. parasites de Vertébrés, Neuchâtel*, 88, 1957; D. A. Humphries, 'The Host-Finding Behaviour of the Hen Flea *Ceratophyllus gallinae* (Schrank) (Siphonaptera)', *Parasitology*, **58**, 403, 1968; D. A. Humphries, 'Behavioural Aspects of the Ecology of the Sand Martin Flea *Ceratophyllus styx jordani* Smit (Siphonaptera)', *Parasitology*, **59**, 311, 1969; S. B. Kendall, 'Relationships Between Species of *Fasciola* and their Molluscan Hosts', *Adv. Parasit.*, **3**, 59, 1965; J. Llewellyn, 'The Life Histories and Population Dynamics of Monogenean Gill Parasites of *Trachurus trachurus* at Plymouth', *J. mar.*

Biol. Ass. U.K., **42**, 587, 1962; R. F. Nigrelli, 'On the Effect of Fish Mucus on *Epibdella melleni* a Monogenetic Trematode of Marine Fishes', *J. Parasit.*, **21** (Supp), 438, 1935; J. P. Thurston, 'The Morphology and Life Cycle of *Protopolystoma xenopi* (Price) Bychowsky in Uganda', *Parasitology*, **54**, 441, 1964; C. A. Wright, 'Relationships Between Trematodes and Molluscs', *Ann. Trop. Med. Parasit.*, **54**, 1, 1960.

8 Host-parasitic relations

A. C. Chandler, 'A Study of the Structure of Feathers with Reference to their Taxonomic Significance', *Univ. Calif. Pub. Zool.*, **13**, 243, 1947; R. T. Damian, 'Molecular Mimicry: Antigen Sharing by Parasite and Host and its Consequences', *Am. Nat.*, **98**, 129, 1964; H. B. N. Hynes and W. L. Nicolas, 'The Development of *Polymorphus minutus* (Goeze, 1782) (Acanthocephala) in the Intermediate Host', *Ann. Trop. Med. Parasit.*, **52**, 376, 1958; E. Mayr and D. Amadon, 'A Classification of Recent Birds', *Am. Mus. Novit.*, No. 1496, 1951; W. P. Rogers and R. I. Sommerville, 'The Infective Stage of Nematode Parasites and its Significance in Parasitism', *Adv. Parasit.*, **1**, 109, 1963; M. Rothschild and B. Ford, 'Breeding of the Rabbit Flea (*Spilopsyllus cuniculi* (Dale) Controlled by the Reproductive Hormones of the Host', *Nature*, **201**, 103, 1964.

9 Parasites and evolution

T. W. N. Cameron, 'Host Specificity and the Evolution of Helminthic Parasites', *Adv. Parasit.*, **2**, 1, 1964; E. C. Dougherty, 'Evolution of Zooparasitic Groups in the Phylum Nematoda, with special reference to Host Distribution', *J. Parasit.*, **37**, 353, 1951; J. Llewellyn, 'The Evolution of Parasitic Platyhelminths', *Symp. Brit. Soc. Parasit.*, **3**, 47, 1965; N. B. Marshall, 'Evolutionary Aspects of Fish Classification', *1er symp. spéc. paras. parasites de Vertébrés, Neuchâtel*, 173, 1957; Clark P. Read, 'Nutrition of Intestinal Helminths', in E. J. L. Soulsby (ed.), *Biology of Parasites*, London and New York, 1966; G. G. Simpson, 'The Principles of Classification and a Classification of Mammals', *Bull. Am. Mus. Nat. Hist.*, **85**, 1945; J. D. Smyth, *The Physiology of Trematodes*, Edinburgh and New York, 1966; J. F. A. Sprent, 'Studies on Ascaroid Nematodes in Pythons: Redefinition of *Ophidascaris filaria* and *Polydelphis anoura*', *Parasitology*, **59**, 129, 1969; Angela Taylor and J. R. Baker, *The Cultivation of Parasites in Vitro*, Oxford and New York, 1968; P. P. Weinstein, 'The In-Vitro Cultivation of Helminths with Reference to Morphogenesis', in E. J. L. Soulsby (ed.), *Biology of Parasites*, London and New York, 1966.

Acknowledgments

Several colleagues have contributed to this book by sending me original material and photographs. I am particularly grateful to Dr J. C. Beaucournu (Rennes) for supplying me with the living flea larvae illustrated in figure 3·1; to Dr Theresa Clay, and the Trustees of the British Museum, for permission to use figures 3·5, 3·6 and 3·7; to M. J. P. L'Hardy (Roscoff), who kindly provided the fine original negatives for figures 2·19 and 2·22; to my friend Dr R. Rausch (Fairbanks) who supplied the material for figure 1·1 and also the original photograph for figure 5·31; to Dr Claude Vaucher, who arranged for the original material at the institute to be photographed; to Dr H. H. Williams (Aberdeen), who provided the original drawing for figure 5·10; and finally to Dr P. C. Young (Commonwealth Bureau of Helminthology), who supplied the material for the photograph reproduced in figure 1·4 (right).

Among my immediate collaborators, I am indebted to Dr Claude Vaucher for reading the manuscript and making various suggestions, and to my secretary, Mme J. Billeter, whose task it was to decipher my handwriting and put together the final version of the manuscript.

Index